# 2016 SQA Past Papers With Answers

# Higher
# BIOLOGY

2014 Specimen Question Paper,
2015 & 2016 Exams

HODDER
GIBSON
AN HACHETTE UK COMPANY

This book contains the official 2014 SQA Specimen Question Paper; 2015 and 2016 Exams for Higher Biology, with associated SQA-approved answers modified from the official marking instructions that accompany the paper.

In addition the book contains study skills advice. This advice has been specially commissioned by Hodder Gibson, and has been written by experienced senior teachers and examiners in line with the new Higher for CfE syllabus and assessment outlines. This is not SQA material but has been devised to provide further guidance for Higher examinations.

Hodder Gibson is grateful to the copyright holders, as credited on the final page of the Answer Section, for permission to use their material. Every effort has been made to trace the copyright holders and to obtain their permission for the use of copyright material. Hodder Gibson will be happy to receive information allowing us to rectify any error or omission in future editions.

Hachette UK's policy is to use papers that are natural, renewable and recyclable products and made from wood grown in sustainable forests. The logging and manufacturing processes are expected to conform to the environmental regulations of the country of origin.

Orders: please contact Bookpoint Ltd, 130 Park Drive, Milton Park, Abingdon, Oxon OX14 4SE. Telephone: (44) 01235 827720. Fax: (44) 01235 400454. Lines are open 9.00–5.00, Monday to Saturday, with a 24-hour message answering service. Visit our website at www.hoddereducation.co.uk. Hodder Gibson can be contacted direct on: Tel: 0141 333 4650; Fax: 0141 404 8188; email: hoddergibson@hodder.co.uk

This collection first published in 2016 by
Hodder Gibson, an imprint of Hodder Education,
An Hachette UK Company
211 St Vincent Street
Glasgow G2 5QY

Typeset by Aptara, Inc.

Printed in the UK

A catalogue record for this title is available from the British Library

ISBN: 978-1-4718-9081-9

3 2 1

2017 2016

# Introduction

## Study Skills – what you need to know to pass exams!

### Pause for thought

Many students might skip quickly through a page like this. After all, we all know how to revise. Do you really though?

*Think about this:*

"IF YOU ALWAYS DO WHAT YOU ALWAYS DO, YOU WILL ALWAYS GET WHAT YOU HAVE ALWAYS GOT."

Do you like the grades you get? Do you want to do better? If you get full marks in your assessment, then that's great! Change nothing! This section is just to help you get that little bit better than you already are.

There are two main parts to the advice on offer here. The first part highlights fairly obvious things but which are also very important. The second part makes suggestions about revision that you might not have thought about but which WILL help you.

### Part 1

DOH! It's so obvious but …

### *Start revising in good time*

Don't leave it until the last minute – this will make you panic.

Make a revision timetable that sets out work time AND play time.

### *Sleep and eat!*

Obvious really, and very helpful. Avoid arguments or stressful things too – even games that wind you up. You need to be fit, awake and focused!

### *Know your place!*

Make sure you know exactly **WHEN and WHERE** your exams are.

### *Know your enemy!*

**Make sure you know what to expect in the exam.**

How is the paper structured?

How much time is there for each question?

What types of question are involved?

Which topics seem to come up time and time again?

Which topics are your strongest and which are your weakest?

Are all topics compulsory or are there choices?

### *Learn by DOING!*

There is no substitute for past papers and practice papers – they are simply essential! Tackling this collection of papers and answers is exactly the right thing to be doing as your exams approach.

### Part 2

People learn in different ways. Some like low light, some bright. Some like early morning, some like evening or night. Some prefer warm, some prefer cold. But everyone uses their BRAIN and the brain works when it is active. Passive learning – sitting gazing at notes – is the most INEFFICIENT way to learn anything. Below you will find tips and ideas for making your revision more effective and maybe even more enjoyable. What follows gets your brain active, and active learning works!

### *Activity 1 – Stop and review*

#### Step 1

When you have done no more than 5 minutes of revision reading STOP!

#### Step 2

Write a heading in your own words which sums up the topic you have been revising.

#### Step 3

Write a summary of what you have revised in no more than two sentences. Don't fool yourself by saying, "I know it, but I cannot put it into words". That just means you don't know it well enough. If you cannot write your summary, revise that section again, knowing that you must write a summary at the end of it. Many of you will have notebooks full of blue/black ink writing. Many of the pages will not be especially attractive or memorable so try to liven them up a bit with colour as you are reviewing and rewriting. **This is a great memory aid, and memory is the most important thing.**

## Activity 2 – Use technology!

Why should everything be written down? Have you thought about "mental" maps, diagrams, cartoons and colour to help you learn? And rather than write down notes, why not record your revision material?

What about having a text message revision session with friends? Keep in touch with them to find out how and what they are revising and share ideas and questions.

Why not make a video diary where you tell the camera what you are doing, what you think you have learned and what you still have to do? No one has to see or hear it, but the process of having to organise your thoughts in a formal way to explain something is a very important learning practice.

Be sure to make use of electronic files. You could begin to summarise your class notes. Your typing might be slow, but it will get faster and the typed notes will be easier to read than the scribbles in your class notes. Try to add different fonts and colours to make your work stand out. You can easily Google relevant pictures, cartoons and diagrams which you can copy and paste to make your work more attractive and **MEMORABLE**.

## Activity 3 – This is it. Do this and you will know lots!

### Step 1

In this task you must be very honest with yourself! Find the SQA syllabus for your subject (www.sqa.org.uk). Look at how it is broken down into main topics called MANDATORY knowledge. That means stuff you MUST know.

### Step 2

BEFORE you do ANY revision on this topic, write a list of everything that you already know about the subject. It might be quite a long list but you only need to write it once. It shows you all the information that is already in your long-term memory so you know what parts you do not need to revise!

### Step 3

Pick a chapter or section from your book or revision notes. Choose a fairly large section or a whole chapter to get the most out of this activity.

With a buddy, use Skype, Facetime, Twitter or any other communication you have, to play the game "If this is the answer, what is the question?". For example, if you are revising Geography and the answer you provide is "meander", your buddy would have to make up a question like "What is the word that describes a feature of a river where it flows slowly and bends often from side to side?".

Make up 10 "answers" based on the content of the chapter or section you are using. Give this to your buddy to solve while you solve theirs.

### Step 4

Construct a wordsearch of at least 10 × 10 squares. You can make it as big as you like but keep it realistic. Work together with a group of friends. Many apps allow you to make wordsearch puzzles online. The words and phrases can go in any direction and phrases can be split. Your puzzle must only contain facts linked to the topic you are revising. Your task is to find 10 bits of information to hide in your puzzle, but you must not repeat information that you used in Step 3. DO NOT show where the words are. Fill up empty squares with random letters. Remember to keep a note of where your answers are hidden but do not show your friends. When you have a complete puzzle, exchange it with a friend to solve each other's puzzle.

### Step 5

Now make up 10 questions (not "answers" this time) based on the same chapter used in the previous two tasks. Again, you must find NEW information that you have not yet used. Now it's getting hard to find that new information! Again, give your questions to a friend to answer.

### Step 6

As you have been doing the puzzles, your brain has been actively searching for new information. Now write a NEW LIST that contains only the new information you have discovered when doing the puzzles. Your new list is the one to look at repeatedly for short bursts over the next few days. Try to remember more and more of it without looking at it. After a few days, you should be able to add words from your second list to your first list as you increase the information in your long-term memory.

## FINALLY! Be inspired...

Make a list of different revision ideas and beside each one write **THINGS I HAVE** tried, **THINGS I WILL** try and **THINGS I MIGHT** try. Don't be scared of trying something new.

And remember – "FAIL TO PREPARE AND PREPARE TO FAIL!"

# Higher Biology

The practice papers in this book give an overall and comprehensive coverage of assessment of **Knowledge** and **Scientific Inquiry** for the new CfE Higher Biology.

We recommend that you download and print a copy of Higher Biology Course Assessment Specification (CAS) pages 8–18 from the SQA website at www.sqa.org.uk.

## The course

The Higher Biology Course consists of three National Units. These are DNA and the Genome, Metabolism and Survival and Sustainability and Interdependence. In each of the units you will be assessed on your ability to demonstrate and apply knowledge of Biology and to demonstrate and apply skills of scientific inquiry. Candidates must also complete an Assignment in which they research a topic in biology and write it up as a report. They also take a Course examination.

## How the course is graded

To achieve a course award for Higher Biology you must pass all three National Unit Assessments which will be assessed by your school or college on a pass or fail basis. The grade you get depends on the following two course assessments, which are set and graded by SQA.

1. An 800–1200 word report based on an Assignment, which is worth 17% of the grade. The Assignment is marked out of 20 marks, with 15 of the marks being for scientific inquiry skills and 5 marks for the application of knowledge.
2. A written course examination is worth the remaining 83% of the grade. The examination is marked out of 100 marks, most of which are for the demonstration and application of knowledge although there are also marks available for skills of scientific inquiry.

This book should help you practice the examination part! To pass Higher Biology with a C grade you will need about 50% of the 120 marks available for the Assignment and the Course Examination combined. For a B you will need roughly 60% and, for an A, roughly 70%.

### The course examination

The Course Examination is a single question paper in two sections.

- The first section is an objective test with 20 multiple choice items for 20 marks.
- The second section is a mix of restricted and extended response questions worth between 2 and 9 marks each for a total of 80 marks. The majority of the marks test knowledge with an emphasis on the application of knowledge. The remainder, test the application of scientific inquiry, analysis and problem solving skills. There will usually be opportunity to comment on or suggest modifications to an experimental situation.

Altogether, there are 100 marks and you will have 2 hours and 30 minutes to complete the paper. The majority of the marks will be straightforward and linked to grade C but some questions are more demanding and are linked to grade A.

## General tips and hints

You should have a copy of the Course Assessment Specification (CAS) for Higher Biology but, if you haven't got one, make sure to download it from the SQA website. This document tells you what can be tested in your examination. It is worth spending some time on this document. This book contains three practice Higher examination papers. One is the SQA specimen paper and there are two past exam papers. Notice how similar they all are in the way in which they are laid out and the types of question they ask – your own course examination is going to be very similar as well, so the value of the papers is obvious! Each paper can be attempted in its entirety or groups of questions on a particular topic or skill area can be attempted. If you are trying a whole examination paper from this book, give yourself 2 hours and 30 minutes maximum to complete it. The questions in each paper are laid out in Unit order. Make sure that you spend time in using the answer section to mark your own work – it is especially useful if you can get someone to help you with this.

The marking instructions give acceptable answers with alternatives. You could even grade your work on an A–D basis. The following hints and tips are related to examination techniques as well as avoiding common mistakes. Remember that if you hit problems with a question, you should ask your teacher for help.

## Section 1

20 multiple-choice items **20 marks**

- Answer on a grid.
- Do not spend more than 30 minutes on this section.
- Some individual questions might take longer to answer than others – this is quite normal and make sure you use scrap paper if a calculation or any working is needed.
- Some questions can be answered instantly– again, this is normal.
- Do not leave blanks – complete the grid for each question as you work through.
- Try to answer each question in your head without looking at the options. If your answer is there you are home and dry!
- If you are not certain, choose the answer that seemed most attractive on first reading the answer options.
- If you are guessing, try to eliminate options before making your guess. If you can eliminate three – you are left with the correct answer even if you do not recognise it!

## Section 2

Restricted and extended response **80 marks**

- Spend about 2 hours on this section.
- Answer on the question paper. Try to write neatly and keep your answers on the support lines if possible – the lines are designed to take the full answer!
- A clue to answer length is the mark allocation – most questions are restricted to 1 mark and the answer can be quite short. If there are 2–4 marks available, your answer will need to be extended and may well have two, three or even four parts.
- The questions are usually laid out in Unit sequence but remember some questions are designed to cover more than one Unit.
- The C-type questions usually start with "State", "Identify", "Give" or "Name" and often need only a word or two in response. They will usually be for 1 mark each.
- Questions that begin with " Explain" and "Describe" are usually A types and are likely to have more than one part to the full answer. You will usually have to write a sentence or two and there may be 2 or even 3 marks available.
- Make sure you read questions over twice before trying to answer – there is often very important information within the question and you are unlikely to be short of time in this examination.

- Using abbreviations like DNA and ATP is fine and the bases of DNA can be given as A, T, G and C. The Higher Biology Course Assessment Specification (CAS) will give you the acceptable abbreviations.
- Don't worry that a few questions are in unfamiliar contexts, that's the idea! Just keep calm and read the questions carefully.
- If a question contains a choice, be sure to spend a minute or two making the best choice for you.
- In experimental questions, you must be aware of what variables are, why controls are needed and how reliability and validity might be improved. It is worth spending time on these ideas – they are essential and will come up year after year.
- Some candidates like to use a highlighter pen to help them focus on the essential points of longer questions – this is a great technique.
- Remember that a conclusion can be seen from data, whereas an explanation will usually require you to supply some background knowledge as well.
- Remember to "use values from the graph" when describing graphical information in words if you are asked to do so.
- Plot graphs carefully and join the plot points using a ruler. Include zeros on your scale where appropriate and use the data table headings for the axes labels.
- Look out for graphs with two Y-axes – these need extra special concentration and anyone can make a mistake!
- If there is a space for calculation given – you will very likely need to use it! A calculator is essential.
- The main types of calculation tend to be ratios, averages, percentages and percentage change – make sure you can do these common calculations.
- Answers to calculations will not usually have more than two decimal places.
- Do not leave blanks. Always have a go, using the language in the question if you can.

## Good luck!

Remember that the rewards for passing Higher Biology are well worth it! Your pass will help you get the future you want for yourself. In the exam, be confident in your own ability. If you're not sure how to answer a question, trust your instincts and just give it a go anyway.

Keep calm and don't panic! GOOD LUCK!

HIGHER

# 2014 Specimen
# Question Paper

National
Qualifications
SPECIMEN ONLY

SQ04/H/02

Biology
Section 1 — Questions

Date — Not applicable

Duration — 2 hours and 30 minutes

Instructions for the completion of Section 1 are given on *Page two* of your question and answer booklet.

Record your answers on the answer grid on *Page three* of your question and answer booklet.

Before leaving the examination room you must give your question and answer booklet to the Invigilator; if you do not, you may lose all the marks for this paper.

**SECTION 1 — 20 marks**

**Attempt ALL questions**

1. The genetic material in human mitochondria is arranged as

   A   linear chromosomes

   B   circular plasmids

   C   circular chromosomes

   D   inner membranes.

2. The main components of a ribosome are

   A   mRNA and tRNA

   B   rRNA and protein

   C   mRNA and protein

   D   rRNA and mRNA.

3. The diagram below represents part of a protein molecule.

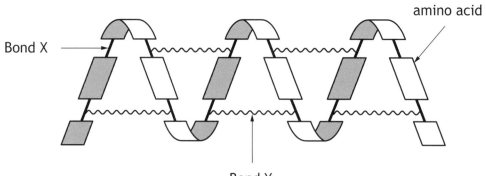

   Which line in the table below identifies bonds X and Y?

|   | Bond X | Bond Y |
|---|--------|--------|
| A | hydrogen | peptide |
| B | hydrogen | hydrogen |
| C | peptide | hydrogen |
| D | peptide | peptide |

4.  Types of single gene mutation are given in the list below.

    1 substitution

    2 insertion

    3 deletion

    Which of these would affect only one amino acid in the polypeptide produced?

    A    1 only

    B    2 only

    C    3 only

    D    2 and 3 only

5.  Which line in the table below describes meristems?

|   | Found in | Type of cell present |
|---|----------|----------------------|
| A | animal | specialised |
| B | animal | unspecialised |
| C | plant | specialised |
| D | plant | unspecialised |

6.  The table below provides information about ancestral and modern Brassica species. The modern species have been produced by hybridisation of two ancestral species followed by a doubling of the chromosome number in the hybrids.

| Brassica species | Ancestral or modern species | Crop | Diploid chromosome number (2 n) |
|---|---|---|---|
| B. oleracea | ancestral | cabbage | 18 |
| B. nigra | ancestral | black mustard | 16 |
| B. rapa | ancestral | turnip | 20 |
| B. juncea | modern | Indian Mustard | 36 |
| B. carinata | modern | Ethiopian Mustard | 34 |
| B. napus | modern | oilseed rape | 38 |

Which of the following shows the ancestral hybridisation and the modern species produced?

A    Cabbage × turnip ⟶ oilseed rape

B    Turnip × black mustard ⟶ Ethiopian mustard

C    Turnip × cabbage ⟶ Indian mustard

D    Cabbage × black mustard ⟶ Indian mustard

7.  The diagram below shows how a molecule might be biosynthesised from building blocks in a metabolic pathway.

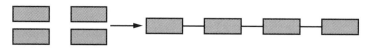

building blocks              biosynthesised molecule

Which line in the table below describes the metabolic process shown in the diagram and the energy relationship involved in the reaction?

|   | Metabolic process | Energy relationship |
|---|---|---|
| A | anabolic | energy used |
| B | anabolic | energy released |
| C | catabolic | energy used |
| D | catabolic | energy released |

8. The graph below shows changes in the α-amylase concentration and the starch content of a barley grain during early growth and development.

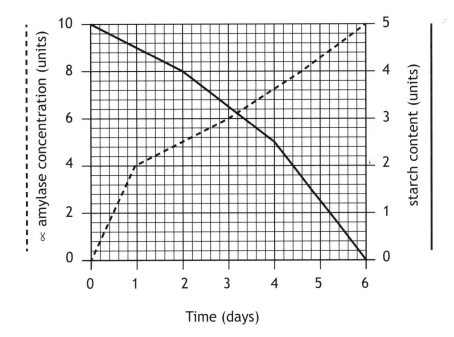

Identify the time by which the starch content of the barley grains had decreased by 50%.

A    2·0 days

B    3·2 days

C    4·0 days

D    6·0 days

9.  The graph below shows the effect of different concentrations of a disinfectant on the number of viable bacteria in liquid culture.

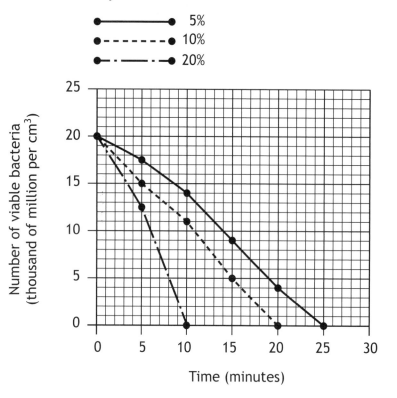

Key: % concentration of disinfectant

What percentage of bacteria was killed by 20% disinfectant after 5 minutes?

A    25

B    37·5

C    62·5

D    75

10.  The diagram below shows a bacterial cell that has been magnified 800 times.

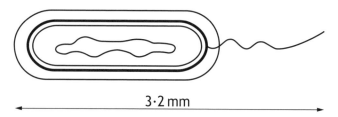

3·2 mm

Calculate the length of the cell in micrometres (μm).

A    0·004

B    0·04

C    0·4

D    4·0

11. The cell membrane contains pumps that actively transport substances.

    Which of the following forms the major component of membrane pumps?

    A    Protein

    B    Phospholipid

    C    Nucleic acid

    D    Carbohydrate

12. Maximum oxygen uptake per kg body mass can be used as a measure of fitness. Four athletes were weighed then given a fitness test during which their maximum oxygen uptake was measured.

    Which line in the table below shows results for the least fit athlete?

| Athlete | Body mass (kg) | Maximum oxygen uptake (litres per minute) |
|---------|----------------|-------------------------------------------|
| A | 60 | 3·6 |
| B | 55 | 3·6 |
| C | 60 | 3·7 |
| D | 55 | 3·7 |

13. The list below gives some adaptations of weed plants.

    1    high seed output

    2    possession of storage organs

    3    vegetative reproduction

    4    long term seed viability

    Which of these are competitive adaptations of annual weeds?

    A    1 and 2 only

    B    1 and 4 only

    C    2 and 3 only

    D    2 and 4 only

14. The table below gives measurements relating to productivity in a field of wheat grown to produce grain for making bread.

| Measurement | Productivity (kg dry mass per hectare per year) |
|---|---|
| plant biomass | 11 250 |
| grain yield | 4500 |

What is the harvest index of this wheat crop?

A    0·4

B    2·5

C    6750

D    15750

15. The action spectrum of photosynthesis is a measure of the ability of plants to

A    absorb all wavelengths of light

B    absorb light of different intensities

C    use light to build up food

D    use light of different wavelengths for photosynthesis.

16. The flow chart below shows the energy flow in a field of potatoes during one year.

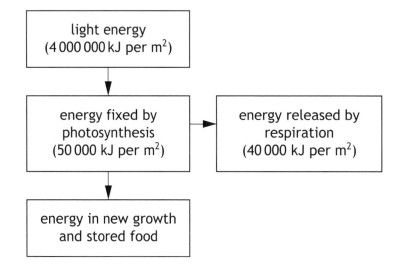

What is the percentage of the available light energy present in new growth and stored food in the potato crop?

A    2·25

B    1·25

C    0·25

D    1·00

**17.** The diagram below represents part of the Calvin cycle within a chloroplast.

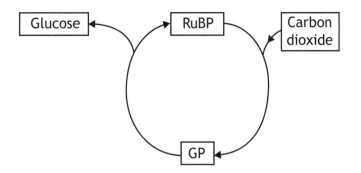

Which line in the table below shows the effect of decreasing $CO_2$ availability on the concentrations of RuBP and GP in the cycle?

|   | RuBP concentration | GP concentration |
|---|---|---|
| A | decrease | decrease |
| B | increase | increase |
| C | decrease | increase |
| D | increase | decrease |

**18.** The list below describes observed behaviour of pigs on a farm.

1   Stereotypic flicking of the head

2   Repeated wounding of other pigs by biting

3   Lying in a position which does not allow suckling

Which of these behaviours indicate poor animal welfare?

A    1 and 2 only

B    1 and 3 only

C    2 and 3 only

D    1, 2 and 3

19. Adult beef tapeworms live in the intestine of humans. Segments of the adult worm are released in the faeces. Embryos that develop from them remain viable for five months. The embryos may be eaten by cattle and develop in their muscle tissue.

Which row in the table below identifies the roles of the human, tapeworm embryo and cattle?

|  | Role | | |
|---|---|---|---|
|  | human | tapeworm embryo | cattle |
| A | host | resistant stage | secondary host |
| B | host | vector | secondary host |
| C | secondary host | vector | host |
| D | secondary host | resistant stage | vector |

20. Ostriches are large birds that live on open plains in Africa. They divide their time between feeding on vegetation and raising their heads to look for predators.

The graphs below show the results of a study on the effect of group size in ostriches on their behaviour.

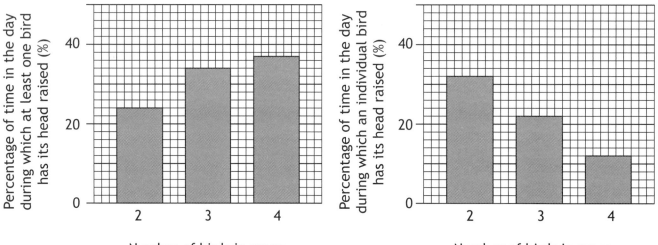

Which of the following is a valid conclusion from these results?

In larger groups, an individual ostrich spends

A    less time with its head raised so the group is less likely to see predators

B    less time with its head raised but the group is more likely to see predators

C    more time with its head raised so the group is more likely to see predators

D    more time with its head raised but the group is less likely to see predators.

**[END OF SECTION 1. NOW ATTEMPT THE QUESTIONS IN SECTION 2 OF YOUR QUESTION AND ANSWER BOOKLET]**

National
Qualifications
SPECIMEN ONLY

Mark

SQ04/H/01

**Biology**
**Section 1 — Answer Grid**
**and Section 2**

Date — Not applicable

Duration — 2 hours and 30 minutes

**Fill in these boxes and read what is printed below.**

Full name of centre

Town

Forename(s)

Surname

Number of seat

Date of birth
Day          Month          Year

D D      M M      Y Y

Scottish candidate number

**Total marks — 100**

**SECTION 1 — 20 marks**

Attempt ALL questions.

Instructions for completion of Section 1 are given on *Page two.*

**SECTION 2 — 80 marks**

Attempt ALL questions.

Write your answers clearly in the spaces provided in this booklet. Additional space for answers and rough work is provided at the end of this booklet. If you use this space you must clearly identify the question number you are attempting. Any rough work must be written in this booklet. You should score through your rough work when you have written your final copy.

Use **blue** or **black** ink.

Before leaving the examination room you must give this booklet to the Invigilator; if you do not you may lose all the marks for this paper.

## SECTION 1— 20 marks

The questions for Section 1 are contained in the question paper SQ04/H/02.
Read these and record your answers on the answer grid on Page three opposite.
Do NOT use gel pens.

1.  The answer to each question is **either** A, B, C or D.  Decide what your answer is, then fill in the appropriate bubble (see sample question below).

2.  There is **only one correct** answer to each question.

3.  Any rough working should be done on the additional space for answers and rough work at the end of this booklet.

**Sample Question**

The thigh bone is called the

    A   humerus

    B   femur

    C   tibia

    D   fibula.

The correct answer is **B**—femur.  The answer **B** bubble has been clearly filled in (see below).

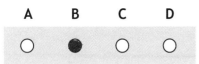

**Changing an answer**

If you decide to change your answer, cancel your first answer by putting a cross through it (see below) and fill in the answer you want.  The answer below has been changed to **D**.

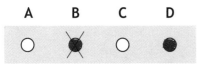

If you then decide to change back to an answer you have already scored out, put a tick (✓) to the **right** of the answer you want, as shown below:

## SECTION 1 — Answer Grid

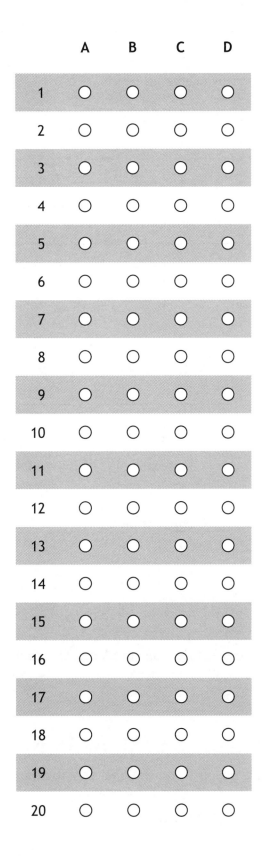

MARKS | DO NOT WRITE IN THIS MARGIN

### SECTION 2 — 80 marks

### Attempt ALL questions

**It should be noted that questions 8 and 14 contain a choice.**

1. The diagram below shows stages in the production of three different proteins that are coded for by one gene.

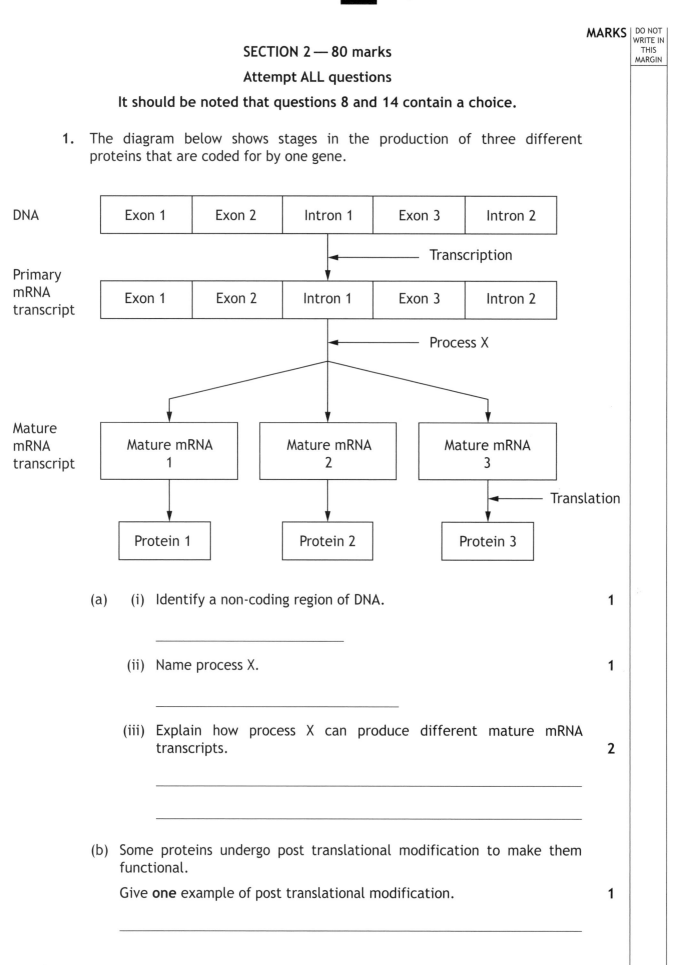

(a) (i) Identify a non-coding region of DNA.    1

_____

(ii) Name process X.    1

_____

(iii) Explain how process X can produce different mature mRNA transcripts.    2

_____

_____

(b) Some proteins undergo post translational modification to make them functional.

Give **one** example of post translational modification.    1

_____

MARKS | DO NOT WRITE IN THIS MARGIN

2.  A chromosome mutation in humans can result in the formation of the Philadelphia chromosome, which is associated with a form of leukaemia.

The stages leading to the formation of a Philadelphia chromosome are shown in the diagram below.

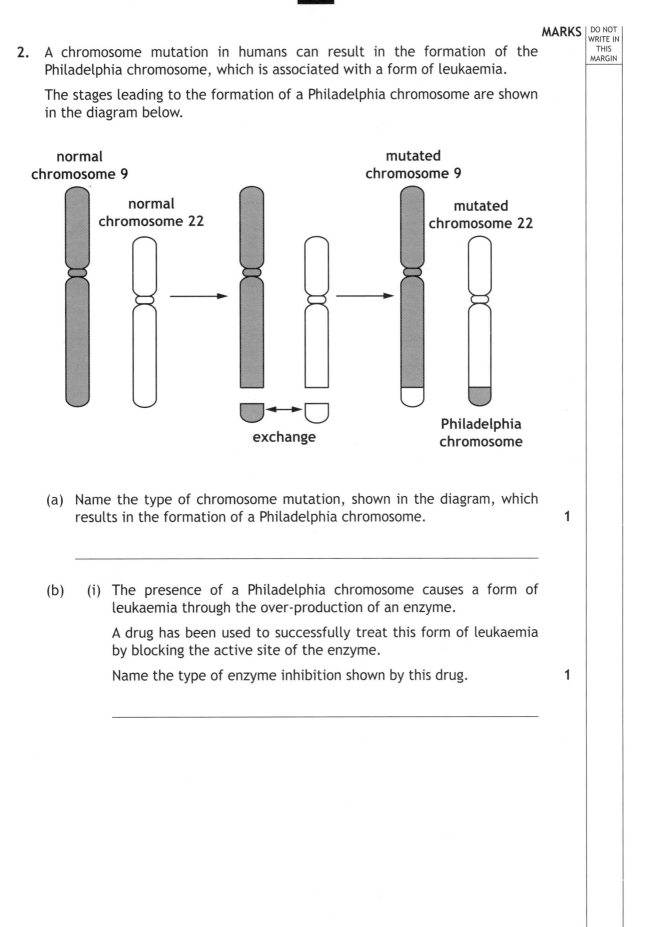

(a)  Name the type of chromosome mutation, shown in the diagram, which results in the formation of a Philadelphia chromosome.

1

(b)  (i)  The presence of a Philadelphia chromosome causes a form of leukaemia through the over-production of an enzyme.

A drug has been used to successfully treat this form of leukaemia by blocking the active site of the enzyme.

Name the type of enzyme inhibition shown by this drug.

1

2. (b) (continued)

(ii) White blood cell counts in humans normally range from 5000 to 10 000 cells per µl of blood.

The table below shows the white blood cell counts from a patient with leukaemia before and after treatment with this drug.

| | Number of white blood cells (per µl blood) |
|---|---|
| Before treatment | 150 000 |
| After treatment | 7500 |

Calculate the percentage decrease in the number of white blood cells after treatment with this drug.

*Space for calculation*

1

_____ %

(iii) Explain how the results suggest that the type of leukaemia in this patient was a result of the presence of a Philadelphia chromosome.

2

_____

_____

_____

MARKS | DO NOT WRITE IN THIS MARGIN

3. The polymerase chain reaction (PCR) amplifies specific sequences of DNA.

The flow chart below shows how a sample of DNA was treated during a cycle of the PCR procedure.

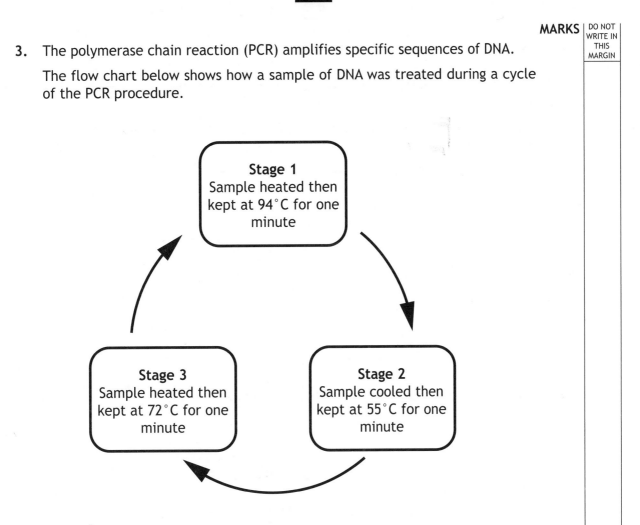

**Stage 1**
Sample heated then kept at 94°C for one minute

**Stage 2**
Sample cooled then kept at 55°C for one minute

**Stage 3**
Sample heated then kept at 72°C for one minute

(a) Explain the purpose of the different heat treatments in Stage 1 and Stage 2.

2

_____

_____

_____

MARKS | DO NOT WRITE IN THIS MARGIN

3. **(continued)**

(b) The number of DNA molecules doubles during each cycle of the PCR procedure.

Calculate the number of cycles needed to produce 128 copies of a single DNA molecule.

1

*Space for calculation*

_____ cycles

(c) The diagram below shows the contents of a tube used in PCR.

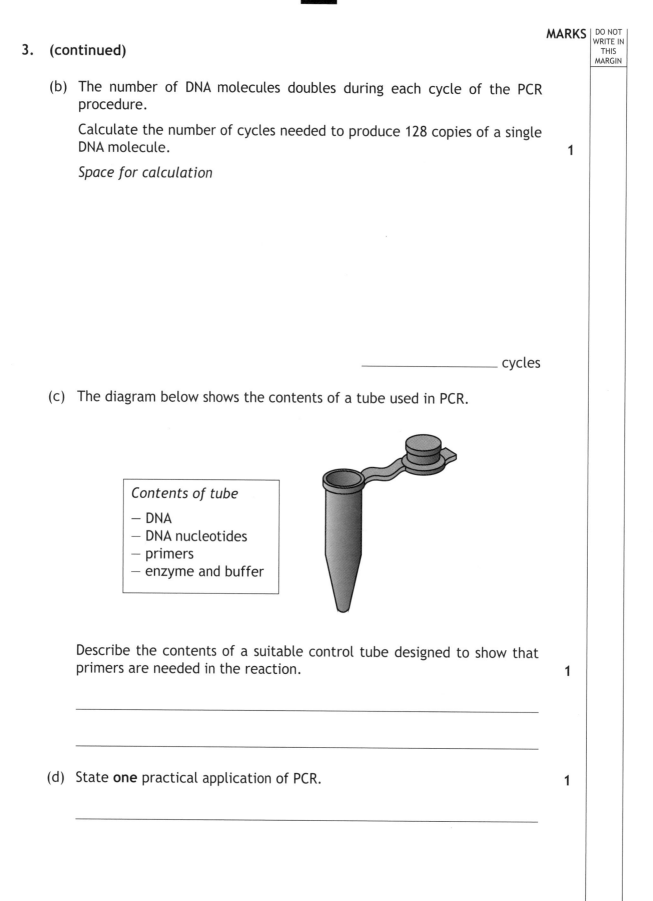

Contents of tube

— DNA
— DNA nucleotides
— primers
— enzyme and buffer

Describe the contents of a suitable control tube designed to show that primers are needed in the reaction.

1

_____

_____

(d) State **one** practical application of PCR.

1

_____

MARKS | DO NOT WRITE IN THIS MARGIN

4. The phylogenetic tree below shows the evolutionary relationship between the three domains of life into which all present day living things can be divided.

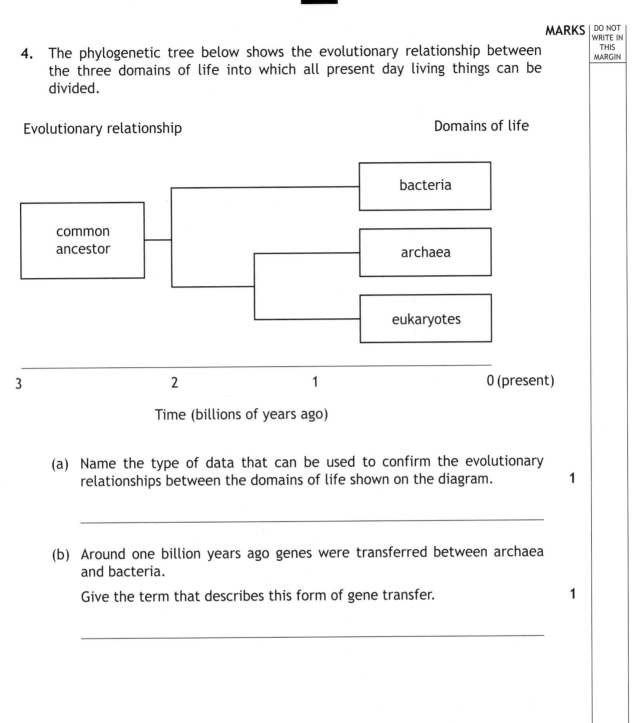

Evolutionary relationship

Domains of life

common ancestor

bacteria

archaea

eukaryotes

3          2          1          0 (present)

Time (billions of years ago)

(a) Name the type of data that can be used to confirm the evolutionary relationships between the domains of life shown on the diagram.    1

(b) Around one billion years ago genes were transferred between archaea and bacteria.

Give the term that describes this form of gene transfer.    1

MARKS | DO NOT WRITE IN THIS MARGIN

4. **(continued)**

(c) The phylogenetic tree below illustrates the evolutionary relationships between primate groups.

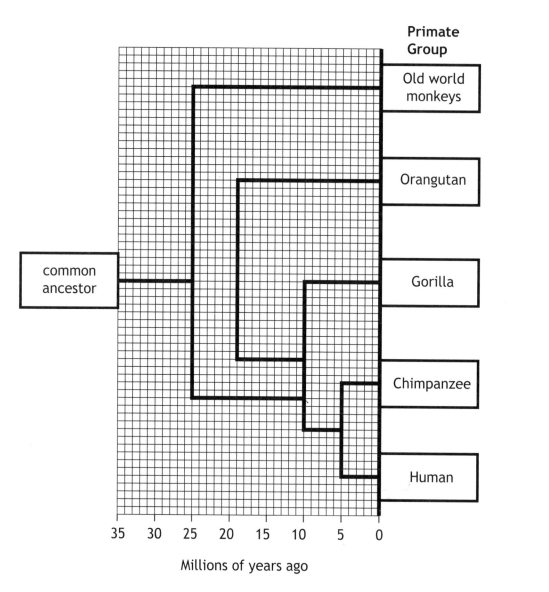

(i) State how long ago the last common ancestor of gorillas and old world monkeys existed.

1

_____ million years ago

MARKS | DO NOT WRITE IN THIS MARGIN

4. **(c)** **(continued)**

(ii) Humans are more closely related to chimpanzees than to orangutans.

Explain how this is known, using information from the phylogenetic tree above.

2

_____

_____

_____

MARKS | DO NOT WRITE IN THIS MARGIN

5.  The diagram below shows some stages in the aerobic respiration of glucose.

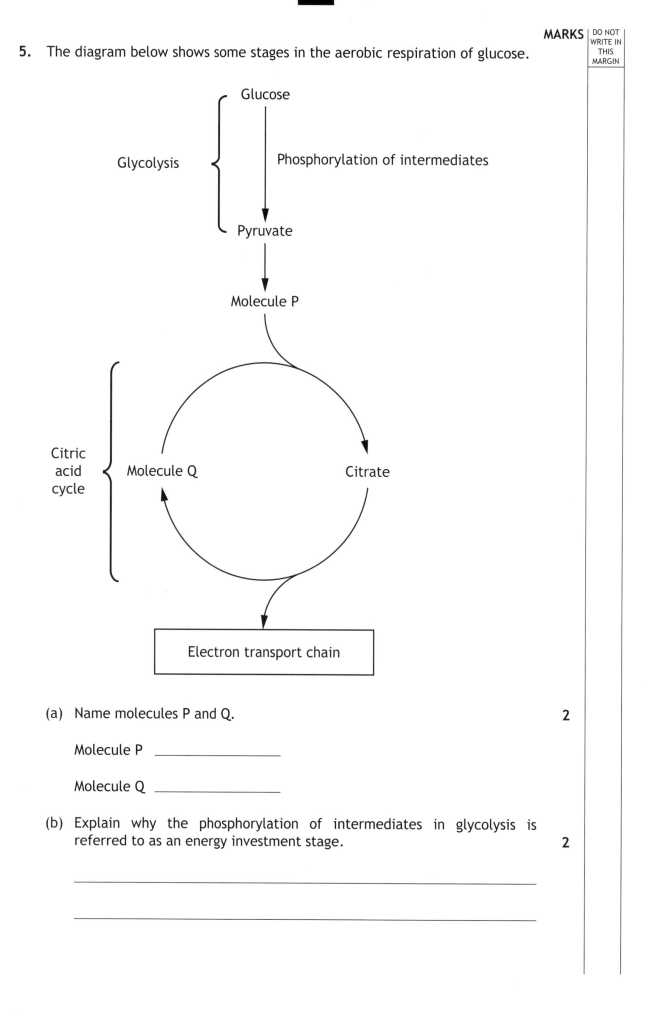

(a)  Name molecules P and Q.                                              2

Molecule P _____

Molecule Q _____

(b)  Explain why the phosphorylation of intermediates in glycolysis is referred to as an energy investment stage.                          2

_____

_____

MARKS | DO NOT WRITE IN THIS MARGIN

5. **(continued)**

(c) Describe the role of the coenzymes NAD and FAD.    2

_____

_____

(d) People who suffer from chronic fatigue syndrome have mitochondria in which some of the proteins embedded in the inner mitochondrial membrane are damaged.

Explain how this might result in the tiredness that is a feature of this condition.    2

_____

_____

_____

MARKS

6. The graph below shows the number of reported cases of hospital acquired infections (HAI) in one hospital between 2002 and 2008. The overall number of patients remained constant during this time.

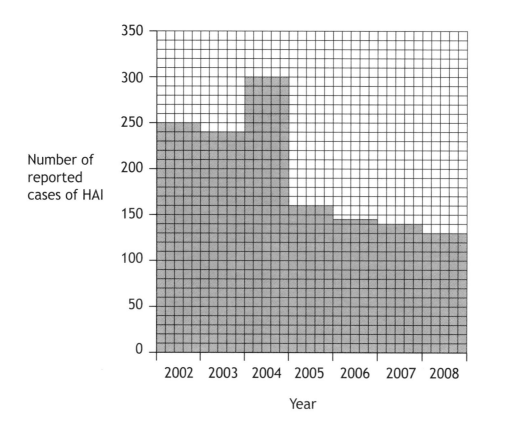

Number of reported cases of HAI

Year

(a) Using information from the graph, calculate the average decrease per year in reported cases of HAI between 2002 and 2008.

*Space for calculation*

1

_____ cases per year

MARKS | DO NOT WRITE IN THIS MARGIN

6. **(continued)**

(b) The decrease in the number of cases in 2005 was due to introduction of a new hand washing procedure at the hospital.

Predict what would happen to the number of reported cases of HAI in 2009.

Circle **one** answer and give a reason for your choice.    1

increase          decrease          stay the same

Reason _____

_____

(c) The table below shows the percentage of cases of HAI in the hospital attributed to two types of bacteria, *Clostridium* and *Staphylococcus*, between 2002 and 2008.

| Bacterial types | Percentage of cases of HAI in each year attributed to bacterial types | | | | | | |
|---|---|---|---|---|---|---|---|
| | 2002 | 2003 | 2004 | 2005 | 2006 | 2007 | 2008 |
| *Clostridium* | 32 | 30 | 30 | 51 | 54 | 57 | 59 |
| *Staphylococcus* | 34 | 32 | 33 | 30 | 31 | 33 | 33 |

Using information in the table, compare the overall trend in the percentage of *Clostridium* cases with that of *Staphylococcus* cases.    2

_____

_____

(d) Using information from the graph and the table, draw a conclusion about the effectiveness of the hand washing procedure against *Staphlycoccus*. Justify your answer.    2

Conclusion _____

_____

Justification _____

_____

6.   (continued)

(e)  Some bacteria form endospores to survive adverse conditions.  Identify which of the two types of bacteria in the table forms endospores and give a reason for your answer.

1

Bacterial type _____

Reason _____

_____

MARKS | DO NOT WRITE IN THIS MARGIN

7. Mammals are regulators and can control their internal environment.

(a) Give **one** reason why it is important for mammals to regulate their body temperature.

1

_____

_____

(b) (i) Name the temperature monitoring centre in the body of a mammal.

1

_____

(ii) State how messages are sent from the temperature monitoring centre to the skin.

1

_____

(c) The blood vessels in the skin of a mammal respond to a decrease in environmental temperature.

(i) Describe this response.

1

_____

(ii) Explain the effect of this response.

1

_____

MARKS | DO NOT WRITE IN THIS MARGIN

8.  Answer **either A or B.**

A   Describe how animals survive adverse conditions.                          4

OR

B   Describe recombinant DNA technology.                                      4

**Labelled diagrams may be used where appropriate.**

MARKS | DO NOT WRITE IN THIS MARGIN

9. The average yield, fat and protein content of the milk from each of three breeds of dairy cattle were determined.

The results are shown in the table below.

| Breed | Average milk yield per cow (kg per day) | Average fat content of milk (%) | Average protein content of milk (%) |
|---|---|---|---|
| Pure bred Holstein | 44·80 | 4·15 | 3·25 |
| F$_1$ hybrid Holstein × Normande | 48·64 | 4·25 | 3·10 |
| F$_1$ hybrid Holstein × Scandinavian Red | 51·52 | 4·25 | 3·15 |

(a) Calculate the percentage increase in average milk yield per cow from the F$_1$ hybrid Holstein × Scandinavian Red compared to pure bred Holstein cattle.    1

*Space for calculation*

_____ %

(b) The fat content of milk is important for butter production.

Calculate the total fat content in the milk produced in a day from a herd of 200 F$_1$ hybrid Holstein × Normande cattle.    1

*Space for calculation*

_____ kg per day

MARKS | DO NOT WRITE IN THIS MARGIN

9.    (continued)

(c)    Select **one** from: average milk yield per cow; average fat content of milk; or average protein content of milk.

For your choice, draw a conclusion about the effects of crossbreeding.    **1**

Choice _____

Conclusion _____

_____

(d)    The development of pure breeds such as Holsteins has led to an accumulation of deleterious recessive alleles.

State the term that describes this.    **1**

_____

(e)    Some $F_2$ offspring from crosses of $F_1$ hybrid Holstein × Scandinavian Red cattle will have less desirable milk-producing characteristics than their parents.

(i)    Give **one** reason for this.    **1**

_____

_____

(ii)    Name a process breeders would have to carry out to maintain the milk-producing characteristics of the $F_1$ hybrids in further generations.    **1**

_____

MARKS | DO NOT WRITE IN THIS MARGIN

10. An investigation was carried out to compare the rate of photosynthesis, at different light intensities, of green algal cells immobilised into gel beads.

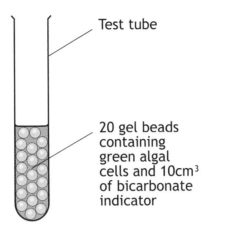

Seven tubes were set up as shown in the diagram and each positioned at a different distance from a light source to alter the light intensity.

Photosynthesis causes the bicarbonate indicator solution to change colour.

After 60 minutes, the bicarbonate indicator solution was transferred from each tube to a colorimeter.

The higher the colorimeter reading, the higher the rate of photosynthesis that has occurred in the tube.

Results are shown in the table.

| Tube | Distance of tube from light source (cm) | Colorimeter reading (units) |
|------|------|------|
| 1 | 25 | 92 |
| 2 | 35 | 92 |
| 3 | 50 | 83 |
| 4 | 75 | 32 |
| 5 | 100 | 14 |
| 6 | 125 | 6 |
| 7 | 200 | 0 |

MARKS | DO NOT WRITE IN THIS MARGIN

10.  **(continued)**

(a)  Identify the dependent variable in this investigation.

1

_____

(b)  Describe how the apparatus could be improved to ensure that temperature was kept constant.

1

_____

(c)  State an advantage of using algae immobilised into gel beads.

1

_____

_____

(d)  Describe how the experimental procedure could be improved to increase the reliability of the results.

1

_____

_____

MARKS | DO NOT WRITE IN THIS MARGIN

**10.** **(continued)**

(e) On the grid below, complete the line graph to show the colorimeter reading against distance of tube from light source.

2

(Additional graph paper if required will be found on *Page twenty-nine*)

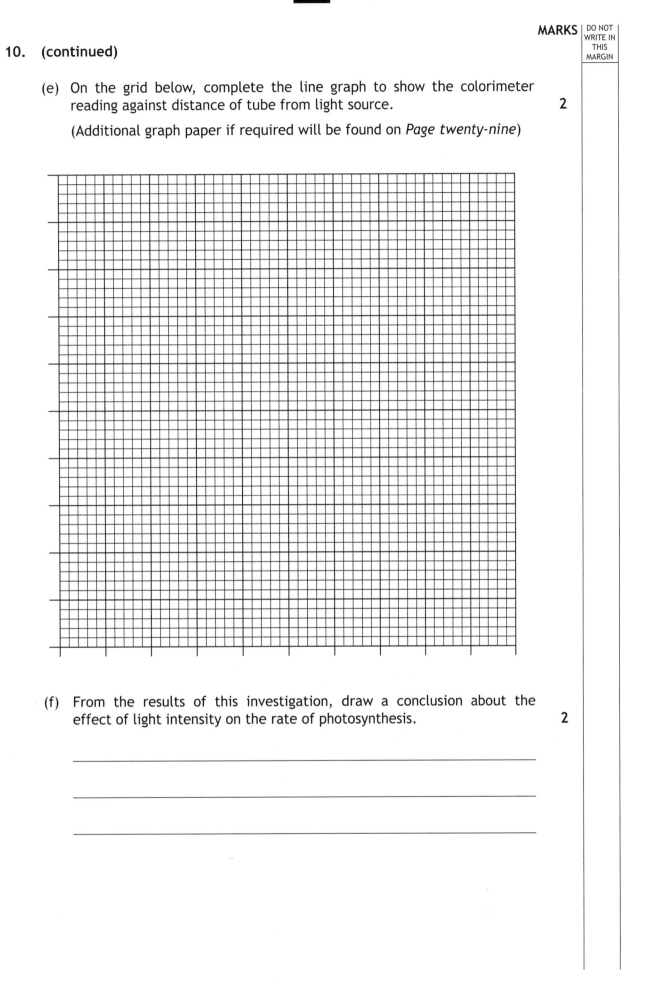

(f) From the results of this investigation, draw a conclusion about the effect of light intensity on the rate of photosynthesis.

2

_____

_____

_____

MARKS | DO NOT WRITE IN THIS MARGIN

**11.** (a) The honey bee (*Apis mellifera*) is a social insect that lives in colonies.

The queen is the only female in a colony that reproduces. Other females are workers that collect food, maintain the colony and care for the developing offspring.

Explain the advantage to the worker bees of caring for the offspring of the queen.

2

_____

_____

(b) The graph below shows the changes in the number of honey bee hives kept by bee-keepers in the USA from 1945 to 2005.

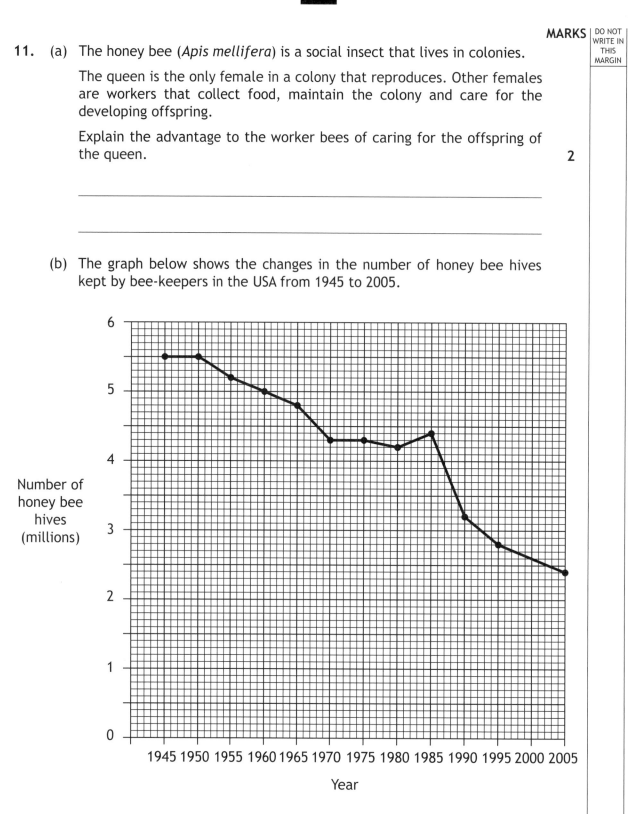

Number of honey bee hives (millions)

Year

MARKS | DO NOT WRITE IN THIS MARGIN

**11.  (b)  (continued)**

(i)  **Using values from the graph,** describe changes in the number of bee hives from 1980 to 1995.

1

_____

_____

(ii)  Calculate the simplest whole number ratio of the number of bee hives in 1965 and 2005.

1

*Space for calculation*

_____ hives in 1965 :  _____ hives in 2005

MARKS | DO NOT WRITE IN THIS MARGIN

**12.** The biodiversity and the genetic diversity of individual species are affected when fragments of woodland become isolated.

The diagram below illustrates habitat fragmentation of an area of woodland over time.

The shaded areas represent woodland.

time

(a) (i) Name **one** component of genetic diversity.  1

_____

(ii) Suggest a reason why a decrease in genetic diversity of an individual species can lead to local extinctions within habitat fragments.  1

_____

_____

(b) Suggest how habitat edge species might affect interior species as the habitat fragments become smaller.  1

_____

_____

(c) Habitat corridors can be created to remedy habitat fragmentation.

(i) State what is meant by the term "habitat corridor".  1

_____

_____

(ii) Explain how a habitat corridor can increase biodiversity after local extinction.  1

_____

_____

MARKS | DO NOT WRITE IN THIS MARGIN

**13.** Japanese knotweed (*Fallopia japonica*) was introduced to Britain as an ornamental plant. It grows to 3 metres in height and has large leaves. It has become naturalised and has colonised many parts of the country where it out-competes native plants.

(a) Give the term used for a naturalised species that eliminates native species.

1

_____

(b) Name **one** resource for which Japanese knotweed may outcompete the native plants.

1

_____

(c) An insect from Japan, which feeds on Japanese knotweed, has been proposed as a biological control agent.

(i) Describe **one** possible risk of introducing this insect into Britain.

1

_____

_____

_____

(ii) Describe a procedure that should be carried out to assess the risk of introducing this insect.

1

_____

_____

MARKS | DO NOT WRITE IN THIS MARGIN

**14.** Answer **either A or B** in the space below.

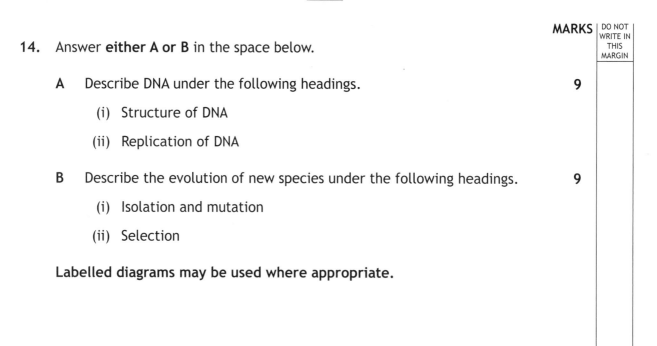

A   Describe DNA under the following headings.                              9

   (i)  Structure of DNA

   (ii) Replication of DNA

B   Describe the evolution of new species under the following headings.    9

   (i)  Isolation and mutation

   (ii) Selection

**Labelled diagrams may be used where appropriate.**

[END OF SPECIMEN QUESTION PAPER]

**ADDITIONAL SPACE FOR ANSWERS AND ROUGH WORK**

ADDITIONAL GRAPH PAPER FOR QUESTION 10 (e)

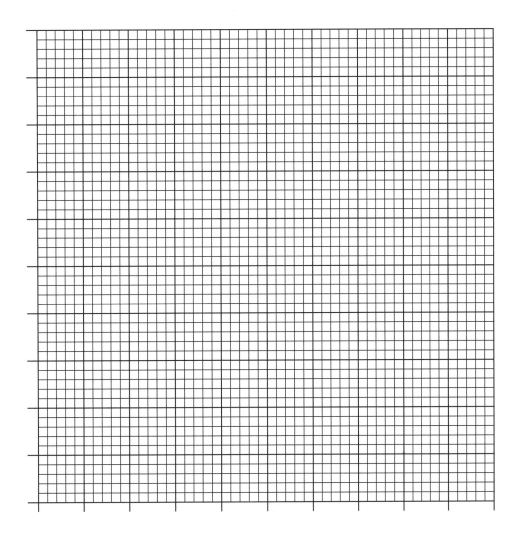

ADDITIONAL SPACE FOR ANSWERS AND ROUGH WORK

*Page thirty*

National
Qualifications
2015

X707/76/02

Biology
Section 1 — Questions

WEDNESDAY, 13 MAY

1:00 PM — 3:30 PM

Instructions for the completion of Section 1 are given on *Page two* of your question and answer booklet X707/76/01.

Record your answers on the answer grid on *Page three* of your question and answer booklet.

Before leaving the examination room you must give your question and answer booklet to the Invigilator; if you do not, you may lose all the marks for this paper.

## SECTION 1 — 20 marks
## Attempt ALL questions

1. Which line in the table below shows features of the human genome?

|   | Contains base sequences that regulate transcription | Contains base sequences transcribed to RNA but never translated | Contains base sequences from which primary transcripts are produced |
|---|---|---|---|
| A | ✗ | ✓ | ✗ |
| B | ✗ | ✗ | ✓ |
| C | ✓ | ✓ | ✗ |
| D | ✓ | ✓ | ✓ |

2. The diagram below shows a eukaryotic gene containing introns and exons and a scale bar representing the number of base pairs in the gene.

Number of base pairs

How many bases will there be in the mature mRNA formed from the primary transcript of this gene?

A    180

B    540

C    560

D    720

3. Which of the following would **not** explain loss of genetic diversity in a population?

A    Inbreeding

B    The founder effect

C    The bottleneck effect

D    No barriers to gene flow

4.  The following are events in the evolution of life on Earth.

    1   Animals appear
    2   Vertebrates appear
    3   Land plants appear

    In which order are these events thought to have occurred?

    A   1 2 3

    B   1 3 2

    C   3 1 2

    D   3 2 1

5.  The graph below shows a molecular clock which compares the amino acid sequences in the protein cytochrome C in various vertebrate groups.

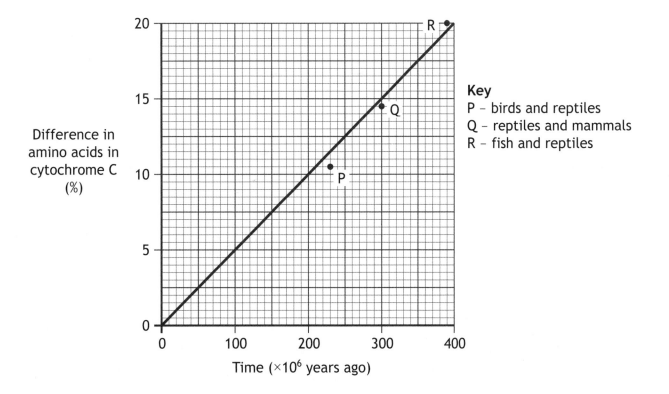

**Key**
P – birds and reptiles
Q – reptiles and mammals
R – fish and reptiles

From the information in the graph, which vertebrate groups shared a common ancestor most recently?

    A   Fish and reptiles

    B   Birds and mammals

    C   Reptiles and mammals

    D   Birds and reptiles

[Turn over

6. The melting temperature of a molecule of DNA ($T_m$) is the temperature at which half of its base pairs separate. $T_m$ is proportional to the percentage of the guanine to cytosine (G–C) base pairs in the molecule as shown on the graph below.

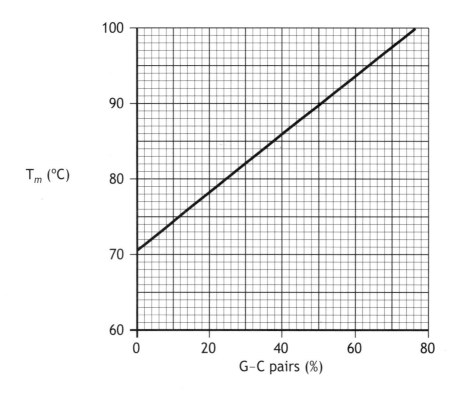

The numbers of base pairs present in a DNA molecule are shown in the table below.

| Number of base pairs present | |
|---|---|
| A–T | G–C |
| 1200 | 800 |

What is $T_m$ for this molecule?

A    78 °C

B    86 °C

C    94 °C

D    96 °C

7. The following are molecules that can be broken down into substrates for respiration.

   1   starch
   2   protein
   3   fat

   Which molecules can be broken down into products which can be converted directly into intermediates of the citric acid cycle?

   A   1 only

   B   1 and 3 only

   C   2 and 3 only

   D   1, 2 and 3

8. The effect of an antibiotic on a bacterial species was tested by spreading a culture of each of the bacterial species on agar plates and adding a disc of absorbent paper soaked in the antibiotic, as shown in the diagram below.

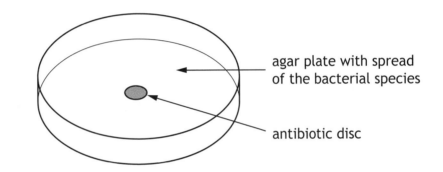

   The plate was incubated for 24 hours at 30 °C and the growth examined.

   Which of the following would be a suitable control for this experiment?

   Repeat the experiment exactly but

   A   with no bacteria

   B   incubate at human body temperature

   C   use a disc with no antibiotic

   D   use a disc with a different antibiotic.

9. Mitochondria are small membrane-bound compartments present in eukaryotic cells.

   One advantage to a mammalian muscle cell of having many small mitochondria is that they provide a

   A   small surface area to volume ratio to increase the uptake of oxygen

   B   large surface area to volume ratio to increase the uptake of oxygen

   C   large surface area to volume ratio to decrease the uptake of carbon dioxide

   D   small surface area to volume ratio to decrease the uptake of carbon dioxide.

10. When salmon migrate from freshwater into seawater, changes in concentration of their surroundings are detected and the activity of the ion pumps in the salmon gills increases. The activity of the ion pumps decrease when the salmon migrate back to freshwater.

Which line in the table below shows the description of the salmon and the control of its ion pumps?

|   | Description of salmon | Control of ion pumps |
|---|---|---|
| A | conformer | by negative feedback |
| B | conformer | behavioural |
| C | regulator | by negative feedback |
| D | regulator | behavioural |

**11.** The rate of sweat production of two individuals, X and Y, was measured during and after a period of exercise.

The results are shown in the graph below.

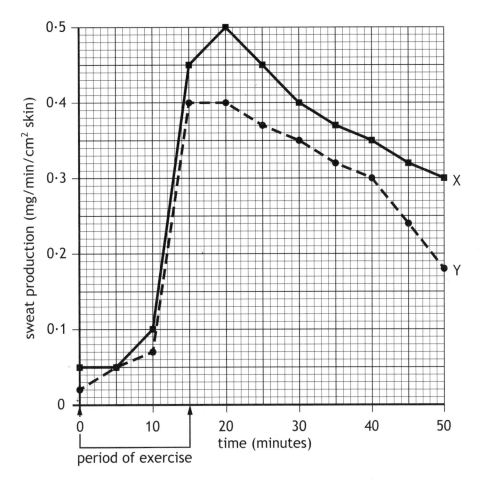

Which of the following conclusions can be drawn from the graph?

A    The rate of sweat production of individual X is always greater than individual Y.

B    Individuals X and Y both reach their maximum sweat production at 20 minutes.

C    Individual X starts increasing sweat production sooner than individual Y.

D    The greatest difference in sweat production by individuals X and Y is at 50 minutes.

**[Turn over**

12.  The table below shows the results of pharmacogenetic tests on a drug designed to treat a liver infection in a group of patients.

| | | Number of patients | |
|---|---|---|---|
| | | beneficial effect on patient | no beneficial effect on patient |
| Number of patients | toxic side-effects | 30 | 15 |
| | no side-effects | 60 | 45 |

What percentage of the patients gained benefit from the drug but showed toxic side-effects?

A   20

B   25

C   30

D   90

13.  The statements below give information on three different bacterial species.

1   *Psychrobacter adeliensis* is found in Antarctica. It has been isolated from coastal ice and grows well at low temperatures.

2   *Thermophilus aquaticus* lives in hot springs and generates ATP by removal of high energy electrons from inorganic molecules.

3   *Escherichia coli* has enzymes with an optimal temperature of 37°C. Most strains of this species are harmless and live in animal intestines although some strains can be harmful to the host animal.

From this information, which of these bacterial species can be classified as extremophile?

A   1 and 2 only

B   1 and 3 only

C   2 only

D   3 only

14. Which of the following results in a transfer of electrons down the electron transport chains during the light dependent reactions of photosynthesis?

    A    NADP is converted to NADPH

    B    Water is split by photolysis

    C    ATP is synthesised

    D    Pigment molecules absorb energy

15. When quantifying plant productivity, the economic yield is the

    A    total biomass produced

    B    biomass of desired product

    C    increase in biomass due to photosynthesis

    D    rate of biomass production per hectare.

16. Soil type is dependent on the composition of its components which in turn affects the productivity of plants growing in it.

    The table below shows the percentage of each component present in four different soils.

|  | Component (%) | | |
|---|---|---|---|
| Soil type | clay | silt | sand |
| sandy clay loam | 20–30 | 0–30 | 50–80 |
| clay loam | 20–35 | 20–60 | 20–50 |
| sandy silt loam | 0–20 | 40–80 | 20–50 |
| silty clay loam | 20–35 | 45–80 | 0–20 |

Which of the following charts represents a clay loam?

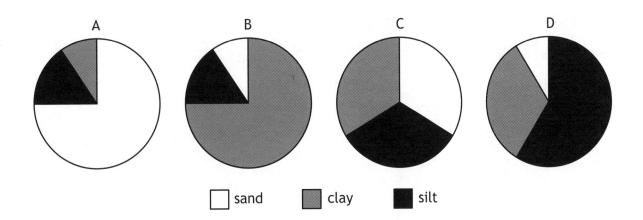

17. The table below shows the number of beet armyworm larvae found in plots of cotton plants.

Some plots were treated with insecticide on 27 June and 1 August and other plots left untreated.

| Sampling date | | Number of beet armyworm larvae | |
|---|---|---|---|
| | | Treated plots | Untreated plots |
| July | 8 | 3 | 3 |
| | 15 | 33 | 2 |
| | 22 | 22 | 17 |
| | 29 | 42 | 10 |
| August | 5 | 120 | 8 |
| | 12 | 160 | 10 |

Which of the following is the most likely explanation for the differences between the treated and untreated plots?

A    The insecticide kills a predator of the larvae

B    The larvae are resistant to the insecticide

C    The beet armyworm breeds in July

D    The larvae have a short lifecycle

18. In primates such as chimpanzees, parental care

A    occurs over a short time period

B    provides time for learning complex social behaviour

C    increases the parent's social status within their group

D    involves appeasement behaviour within a group.

19. Altruistic behaviour between closely related animals

A    reduces competition between individuals in the population

B    increases the survival chances of the donor animal

C    increases the frequency of shared genes in the next generation

D    reduces unnecessary aggression and conflict in social groups.

20. A species that plays a role vital for the survival of many other species in an ecosystem is called

   A   a keystone species

   B   a native species

   C   an invasive species

   D   a dominant species.

**[END OF SECTION 1. NOW ATTEMPT THE QUESTIONS IN SECTION 2
OF YOUR QUESTION AND ANSWER BOOKLET]**

[BLANK PAGE]

DO NOT WRITE ON THIS PAGE

# H

National
Qualifications
2015

Mark

X707/76/01

## Biology
### Section 1 — Answer Grid
### and Section 2

WEDNESDAY, 13 MAY

1:00 PM — 3:30 PM

---

**Fill in these boxes and read what is printed below.**

Full name of centre

Town

Forename(s)

Surname

Number of seat

Date of birth

Day    Month    Year

Scottish candidate number

**Total marks — 100**

**SECTION 1 — 20 marks**

Attempt ALL questions.

Instructions for completion of Section 1 are given on *Page two*.

**SECTION 2 — 80 marks**

Attempt ALL questions.

Questions 5 and 13 each contain a choice.

Write your answers clearly in the spaces provided in this booklet. Additional space for answers and rough work is provided at the end of this booklet. If you use this space you must clearly identify the question number you are attempting. Any rough work must be written in this booklet. You should score through your rough work when you have written your final copy.

Use **blue** or **black** ink.

Before leaving the examination room you must give this booklet to the Invigilator; if you do not you may lose all the marks for this paper.

## SECTION 1— 20 marks

The questions for Section 1 are contained in the question paper X707/76/02.
Read these and record your answers on the answer grid on *Page three* opposite.
Use **blue** or **black** ink. Do NOT use gel pens or pencil.

1. The answer to each question is **either** A, B, C or D. Decide what your answer is, then fill in the appropriate bubble (see sample question below).

2. There is **only one correct** answer to each question.

3. Any rough working should be done on the additional space for answers and rough work at the end of this booklet.

### Sample Question

The thigh bone is called the

    A   humerus

    B   femur

    C   tibia

    D   fibula.

The correct answer is **B**—femur. The answer **B** bubble has been clearly filled in (see below).

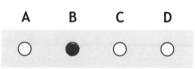

### Changing an answer

If you decide to change your answer, cancel your first answer by putting a cross through it (see below) and fill in the answer you want. The answer below has been changed to **D**.

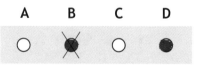

If you then decide to change back to an answer you have already scored out, put a tick (✓) to the **right** of the answer you want, as shown below:

          or

## SECTION 1 — Answer Grid

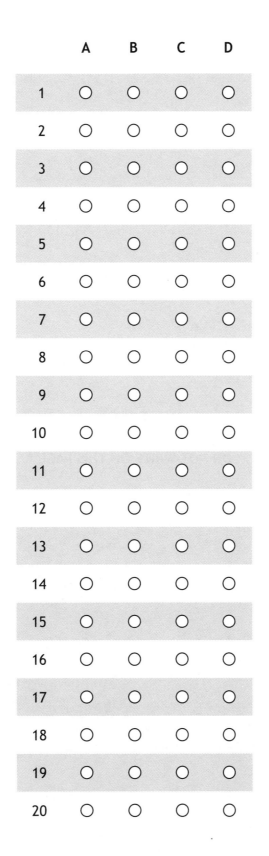

[BLANK PAGE]

DO NOT WRITE ON THIS PAGE

[Turn over for Question 1 on *Page six*

**DO NOT WRITE ON THIS PAGE**

MARKS | DO NOT WRITE IN THIS MARGIN

**SECTION 2 — 80 marks**

**Attempt ALL questions**

**Questions 5 and 13 each contain a choice.**

1. Sauerkraut is a food produced by preserving cabbage. Preservation involves inhibition of the bacteria which can spoil the food. *Lactobacillus* is anaerobic and, unlike most bacteria, grows well at low pH.

   The diagram below shows stages in fermentation of the glucose in cabbage by *Lactobacillus*.

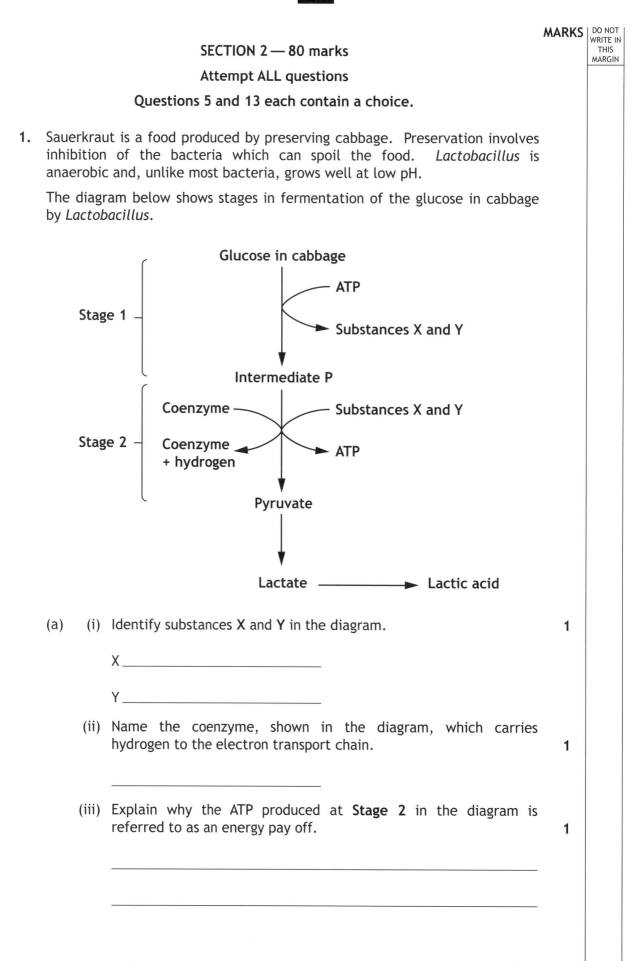

   (a) (i) Identify substances **X** and **Y** in the diagram.  1

   X _____

   Y _____

   (ii) Name the coenzyme, shown in the diagram, which carries hydrogen to the electron transport chain.  1

   _____

   (iii) Explain why the ATP produced at **Stage 2** in the diagram is referred to as an energy pay off.  1

   _____

   _____

MARKS | DO NOT WRITE IN THIS MARGIN

1.  **(continued)**

(b)  The flow chart below shows how cabbage can be processed to produce sauerkraut.

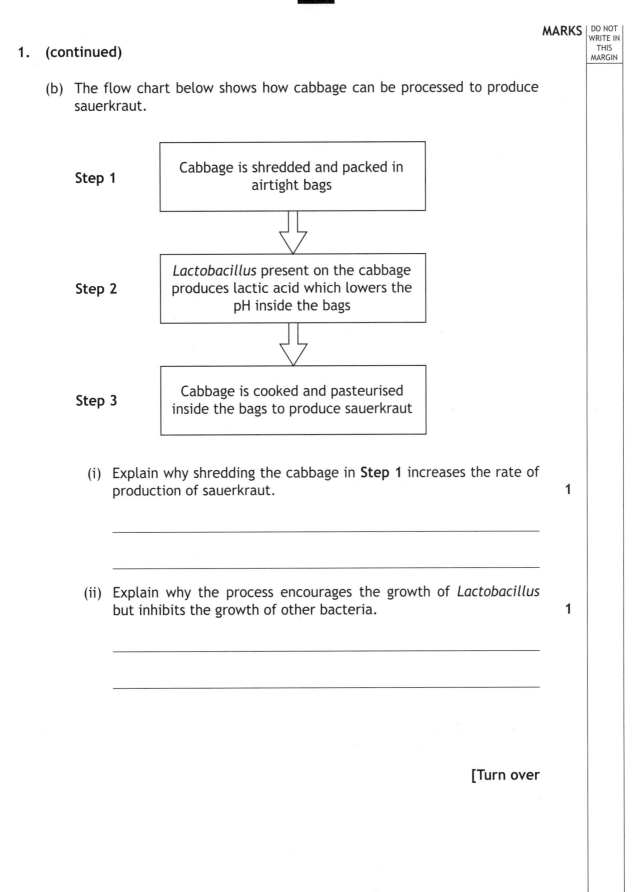

Step 1 — Cabbage is shredded and packed in airtight bags

Step 2 — *Lactobacillus* present on the cabbage produces lactic acid which lowers the pH inside the bags

Step 3 — Cabbage is cooked and pasteurised inside the bags to produce sauerkraut

(i)  Explain why shredding the cabbage in **Step 1** increases the rate of production of sauerkraut.    1

_____

_____

(ii)  Explain why the process encourages the growth of *Lactobacillus* but inhibits the growth of other bacteria.    1

_____

_____

[Turn over

*Page seven*

MARKS | DO NOT WRITE IN THIS MARGIN

2. Erythropoietin (EPO) is a protein synthesised in the kidneys which is involved in red blood cell production. Some individuals with kidney disease have low red blood cell counts and can be treated with EPO.

EPO can be produced by recombinant DNA technology in which the human EPO gene was inserted into a specially prepared bacterial plasmid.

The diagram below shows the prepared bacterial plasmid before and after it was modified by the insertion of a human EPO gene.

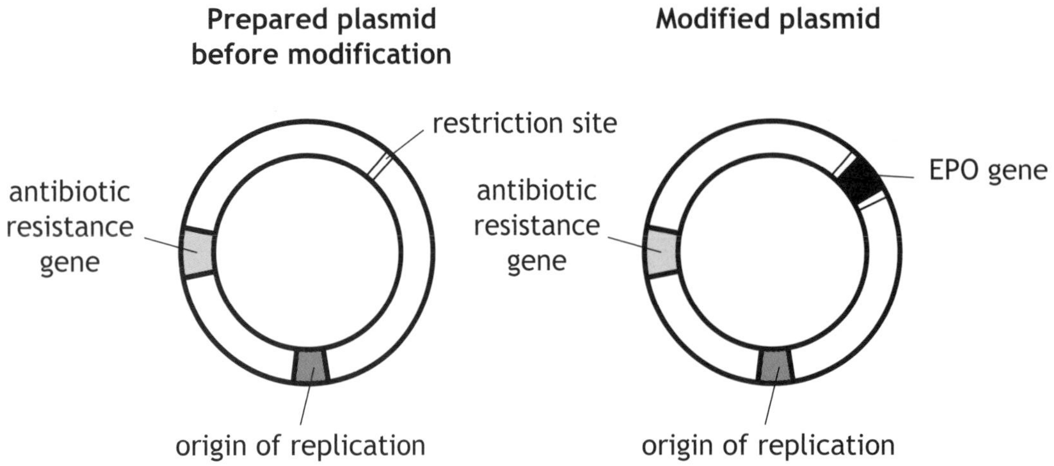

**Prepared plasmid before modification**

**Modified plasmid**

restriction site

antibiotic resistance gene

antibiotic resistance gene

EPO gene

origin of replication

origin of replication

(a) Explain the importance of removing the EPO gene from a human chromosome with the **same** restriction endonuclease that was used to open the bacterial plasmid.

1

_____

_____

(b) Name the enzyme used to seal the EPO gene into the bacterial plasmid.

1

_____

(c) Modified plasmids were mixed with bacteria. Some bacterial cells were transformed by taking up the modified plasmids but others were not.

Use information from the diagram to suggest how a culture containing only the transformed bacteria was obtained.

1

_____

_____

(d) Identify the section of the modified plasmid shown in the diagram which ensured that it could be copied and passed to daughter cells when transformed bacteria divided.

1

_____

MARKS | DO NOT WRITE IN THIS MARGIN

2.  (continued)

(e)  The EPO protein produced by the transformed bacteria is inactive.

(i)  Suggest a reason why bacteria produce EPO protein which is inactive.

1

_____

_____

(ii)  Suggest how recombinant DNA technology could be used to produce an active form of the EPO protein.

1

_____

_____

[Turn over

MARKS | DO NOT WRITE IN THIS MARGIN

3. (a) The yeast *Kluyveromyces marxianus* uses lactose as a respiratory substrate. An investigation was carried out into the effect of lactose concentration on ethanol production by this yeast species. Five flasks were set up each containing $5\,cm^3$ of yeast suspension and $100\,cm^3$ of 4, 8, 12, 16 or 20% lactose solution. The flasks were sealed to maintain anaerobic conditions.

Samples were removed from each flask at 12 and 36 hours and the concentration of ethanol was determined. Results are shown in the table below.

| Lactose concentration (%) | Ethanol concentration (g per 100 cm³) | |
|---|---|---|
| | *12 hours* | *36 hours* |
| 4 | 1·20 | 1.65 |
| 8 | 1·55 | 2·80 |
| 12 | 2·00 | 4·25 |
| 16 | 2·80 | 3·25 |
| 20 | 2·80 | 6·50 |

(i) Identify the independent variable.

1

_____

(ii) Identify **one** variable not already mentioned that should be kept constant so that a valid conclusion can be drawn.

1

_____

(b) Describe the relationship between the lactose concentration and ethanol concentration at 12 hours growth.

2

_____

_____

(c) Calculate the percentage increase in ethanol concentration between 12 and 36 hours growth in the 4% lactose flask.

1

*Space for calculation*

_____ %

MARKS | DO NOT WRITE IN THIS MARGIN

3.  (continued)

(d) Air leaked into the 16% lactose flask between 12 and 36 hours growth. Explain why this resulted than a lower than expected ethanol concentration.

1

_____

_____

_____

[Turn over

MARKS | DO NOT WRITE IN THIS MARGIN

4. The northern blossom bat *Macroglossus minimus* is an Asian species which has a high metabolic rate and a daily rhythm of torpor.

The metabolic rates and body temperatures of a group of these bats were recorded every four hours over a 24 hour cycle and the results are shown on the graph below.

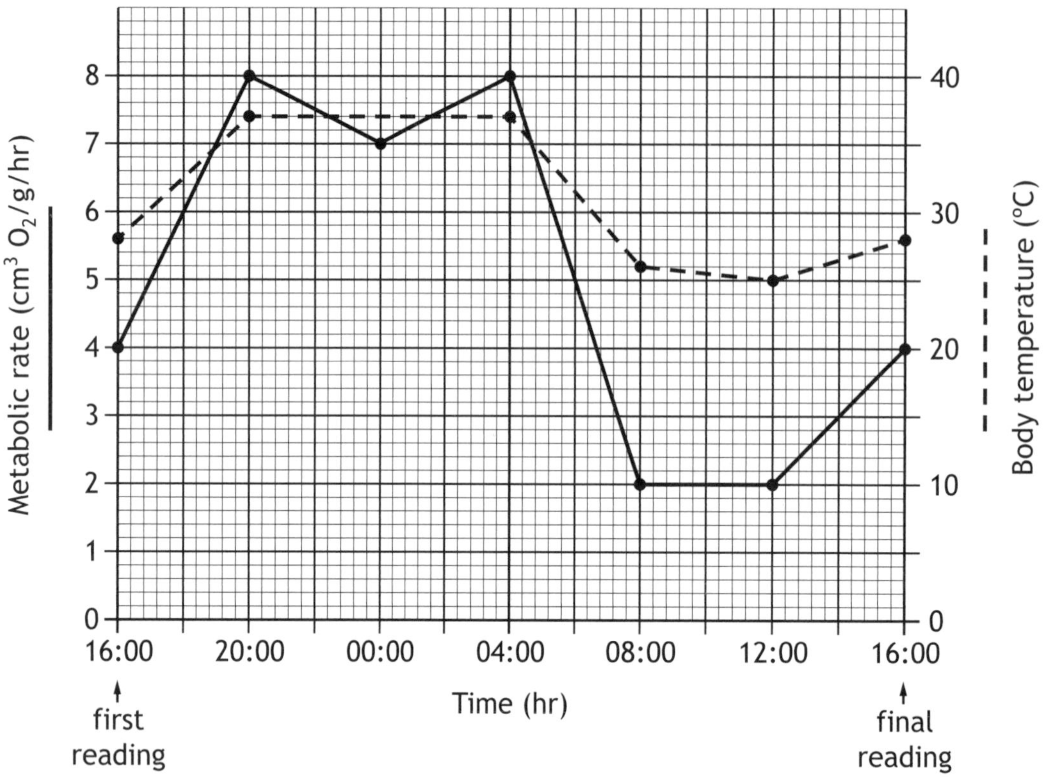

(a) Calculate the oxygen consumption of a 16 g bat at 00:00 hours.

*Space for calculation*

1

_____ cm³ O₂ per hr

(b) Tick (✓) **one** box to identify the period when the bats were in full torpor and justify your answer.

2

16:00 – 20:00     20:00 – 00:00     04:00 – 08:00     08:00 – 12:00

☐               ☐               ☐               ☐

Justification _____

_____

MARKS | DO NOT WRITE IN THIS MARGIN

4.  **(continued)**

(c) Give **one** benefit to bats of their daily torpor.  1

_____

_____

(d) Blossom bats are nocturnal.

Give **one other** behavioural adaptation of animals with high metabolic rates to allow survival in adverse conditions.  1

_____

_____

[Turn over

MARKS | DO NOT WRITE IN THIS MARGIN

5.  Answer **either A or B** in the space below.

    A   Describe the arrangement of heart chambers in birds and amphibians
        and relate this to their metabolic rates.                                4

    OR

    B   Describe competitive and non-competitive inhibition of enzyme action.     4

[Turn over for Question 6 on *Page sixteen*

**DO NOT WRITE ON THIS PAGE**

MARKS | DO NOT WRITE IN THIS MARGIN

6. The diagram below shows some stages in the Calvin cycle of photosynthesis.

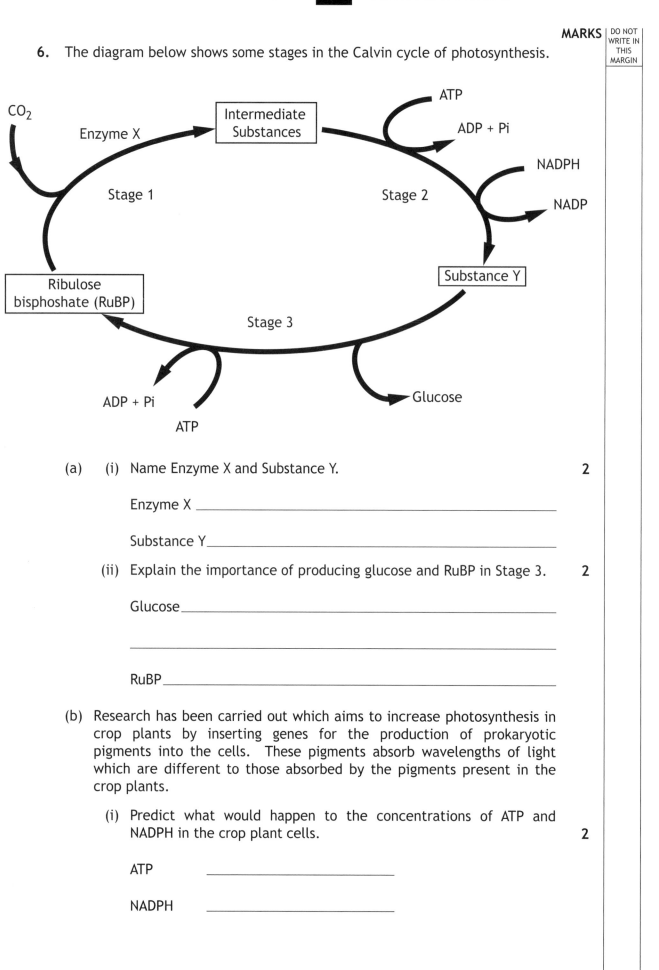

(a) (i) Name Enzyme X and Substance Y. 2

Enzyme X _____

Substance Y_____

(ii) Explain the importance of producing glucose and RuBP in Stage 3. 2

Glucose_____

_____

RuBP_____

(b) Research has been carried out which aims to increase photosynthesis in crop plants by inserting genes for the production of prokaryotic pigments into the cells. These pigments absorb wavelengths of light which are different to those absorbed by the pigments present in the crop plants.

(i) Predict what would happen to the concentrations of ATP and NADPH in the crop plant cells. 2

ATP          _____

NADPH       _____

MARKS | DO NOT WRITE IN THIS MARGIN

6.  (b)  (continued)

(ii) Genetically modified (GM) crops are evaluated in field trials.

Certain experimental procedures are required when setting up field trials to compare GM and non GM crops.

Give **one** such procedure and explain how it allows valid conclusions to be drawn.

2

Procedure _____

_____

Explanation _____

_____

[Turn over

MARKS | DO NOT WRITE IN THIS MARGIN

7. The parasite *Schistosoma mansoni* causes the condition schistosomiasis in humans.

The condition is common in tropical regions where the parasite is often present in fresh water. Humans can be infected if they enter water containing the parasite.

The life cycle of *Schistosoma mansoni* is shown below.

```
                    ┌─────────────────────────┐
                    │ Free swimming immature  │
                    │ parasites enter the human│
                    │ body through the skin   │
                    └─────────────────────────┘
       ┌──────────────────────┐        ┌──────────────────────┐
       │ After further         │        │ They develop into     │
       │ development,          │        │ mature                │
       │ free swimming immature│        │ parasites and migrate │
       │ parasites are released│        │ to                    │
       │ from the snails into  │        │ the intestines and    │
       │ fresh water           │        │ bladder               │
       └──────────────────────┘        └──────────────────────┘
       ┌──────────────────────┐        ┌──────────────────────┐
       │ Immature parasites    │        │ Female parasites      │
       │ enter the tissues of  │        │ produce eggs which are│
       │ fresh water snails    │        │ released in human     │
       │                       │        │ faeces and urine      │
       └──────────────────────┘        └──────────────────────┘
                    ┌─────────────────────────┐
                    │ The eggs hatch if the   │
                    │ faeces or urine reaches │
                    │ fresh water             │
                    └─────────────────────────┘
```

(a) Explain why *Schistosoma mansoni* is described as a parasite.    1

_____

(b) Identify the secondary host and suggest a benefit to *Schistosoma mansoni* of including a secondary host in its life cycle.    2

Secondary host _____

Benefit _____

_____

(c) Describe **one** measure which could be adopted to reduce the number of cases of schistosomiasis.    1

_____

MARKS | DO NOT WRITE IN THIS MARGIN

8. Harlequin ladybirds, *Harmonia axyridis*, were introduced to the UK from their native habitat in Eastern Asia in order to reduce the population of aphids, which feed on crop plants.

Since their introduction, harlequin ladybirds have spread rapidly and their population has dramatically increased. As a result the populations of some ladybird species have dramatically decreased, although the population of native seven-spot ladybirds has remained relatively stable.

(a) Name this control method used to manage the population of aphids.

    **1**

_____

(b) Using the information given, explain why the harlequin ladybird can be described as an invasive species.

    **1**

_____

_____

(c) Suggest one reason why the population size of the seven-spot ladybird has remained relatively stable.

    **1**

_____

(d) Give a reason why the population of harlequin ladybirds has increased more quickly in the UK than in their native habitat.

    **1**

_____

_____

[Turn over

MARKS | DO NOT WRITE IN THIS MARGIN

9. Alfalfa is a crop plant often grown for cattle food.

In a field trial, alfalfa was grown in six plots each of which had been treated with a different level of phosphate fertiliser. The alfalfa was harvested after 24 weeks of growth and the total dry mass of the crop at each fertiliser level was calculated. The protein content of the alfalfa grown at each fertiliser level was determined.

The results are shown in the graph below.

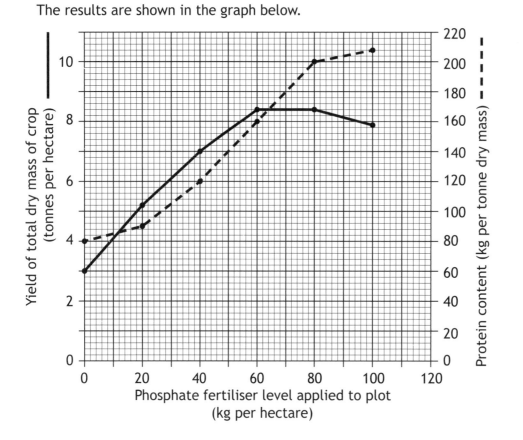

(a)  (i)  **Use values from the graph** to describe the changes in the yield of total dry mass of the crop as the phosphate fertiliser level was increased from 0 to 100 kg per hectare.

2

_____

_____

_____

_____

(ii) Predict the protein content of an alfalfa crop if 120 kg of phosphate fertiliser per hectare had been applied.

1

_____ kg per tonne dry mass

MARKS | DO NOT WRITE IN THIS MARGIN

**9. (a) (continued)**

(iii) Calculate the total mass of protein produced from one hectare when 40 kg of phosphate fertiliser per hectare was applied.

*Space for calculation*

1

_____ kg

(b) In a feeding trial, three groups of 10 cattle were fed with alfalfa of different protein contents over a 25 day period. The cattle were weighed at the beginning and end of this period and the average increase in their body mass calculated.

The results are shown in the table below.

| Cattle group | Protein content of alfalfa fed to cattle (kg per tonne dry mass) | Average increase in body mass of cattle over a 25 day period (kg) |
|---|---|---|
| 1 | 80 | 12 |
| 2 | 90 | 15 |
| 3 | 120 | 17 |

(i) State how the design of the feeding trial ensured the reliability of the results.

1

_____

(ii) Using the information from the **table**, calculate the average increase in body mass per day of the cattle in Group 2.

*Space for calculation*

1

_____ kg per day

(iii) Using information from the **graph and table**;

1    suggest the phosphate fertiliser level which was applied in the production of the alfalfa which the cattle in Group 2 were fed;

1

_____ kg per hectare

2    draw a conclusion about how phosphate fertiliser levels applied to the alfalfa affected the growth of cattle in the feeding trial.

1

_____

MARKS | DO NOT WRITE IN THIS MARGIN

9. **(continued)**

(c) In terms of food security, explain why using agricultural land to grow cereal for human consumption rather than to grow cattle food would produce more food for humans per unit area.

1

_____

_____

[Turn over for Question 10 on *Page twenty-four*

**DO NOT WRITE ON THIS PAGE**

MARKS | DO NOT WRITE IN THIS MARGIN

10. *Staphylococcus aureus (S.aureus)* is a species of bacteria that lives on human skin. This species of bacteria can cause infections if it enters the body through a wound. *S.aureus* infections can be treated with antibiotics such as methicillin and penicillin.

Infections can be caused by a strain of *S.aureus* called MRSA which is resistant to methicillin and penicillin and is becoming more common.

(a) The MRSA strain has developed resistance to antibiotics by gene transfer from another organism.

Identify the correct statement(s) relating to MRSA antibiotic resistance.

Tick (✓) the correct box(es). 　　　　　2

| | |
|---|---|
| MRSA has developed antibiotic resistance through **horizontal** gene transfer from another organism. | |
| MRSA has developed antibiotic resistance through **vertical** gene transfer from another organism. | |
| This type of gene transfer in bacteria brings about a **rapid** evolutionary change. | |
| This type of gene transfer in bacteria brings about a **slow** evolutionary change. | |

(b) Explain how the overuse of antibiotics has led to the increase in the population of MRSA. 　　　　　2

_____

_____

_____

MARKS | DO NOT WRITE IN THIS MARGIN

10.    (continued)

(c)  Samples were taken from a patient suspected of having a bacterial infection.  The samples were used to inoculate plates of agar as shown in the diagram below.

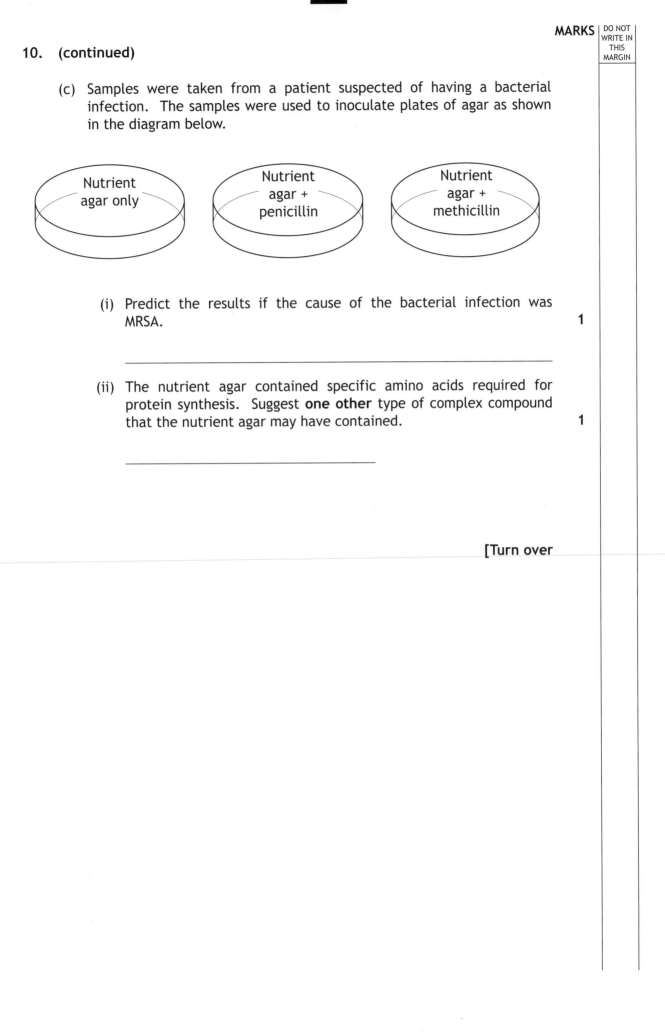

(i)  Predict the results if the cause of the bacterial infection was MRSA.

1

_____

(ii)  The nutrient agar contained specific amino acids required for protein synthesis.  Suggest **one other** type of complex compound that the nutrient agar may have contained.

1

_____

[Turn over

MARKS | DO NOT WRITE IN THIS MARGIN

11. Patients requiring an organ transplant are tissue typed to match with potential donors. Polymerase chain reaction (PCR) and gel electrophoresis are used to compare DNA sequences of the patient with those of donors. Gel electrophoresis separates mixtures of DNA fragments according to size. The presence of a specific DNA band indicates that a donor is a suitable match.

Patient and potential donor samples were compared with a DNA ladder.

The DNA ladder contains fragments of DNA, separated by gel electrophoresis, which are of a known size and measured in base pairs (bp). The distances the DNA fragments travelled were measured and are shown in the table below. The diagram below shows the result of the gel electrophoresis.

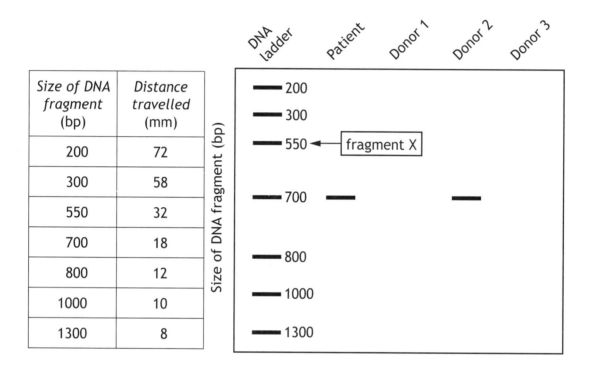

| Size of DNA fragment (bp) | Distance travelled (mm) |
|---|---|
| 200 | 72 |
| 300 | 58 |
| 550 | 32 |
| 700 | 18 |
| 800 | 12 |
| 1000 | 10 |
| 1300 | 8 |

(a) The gel used for electrophoresis contains agarose. Calculate the mass of agarose required to make 30 cm³ of a 0·8% agarose gel.

*Space for calculation*

_____ g

1

(b) Using information in the **table** and the **diagram** give the distance travelled by fragment X in the DNA ladder.

_____ mm

1

MARKS | DO NOT WRITE IN THIS MARGIN

**11. (continued)**

(c) On the grid below, draw a line graph to show the distance travelled against the size of DNA fragment.

2

(Additional graph paper if required will be found on *Page thirty-three*.)

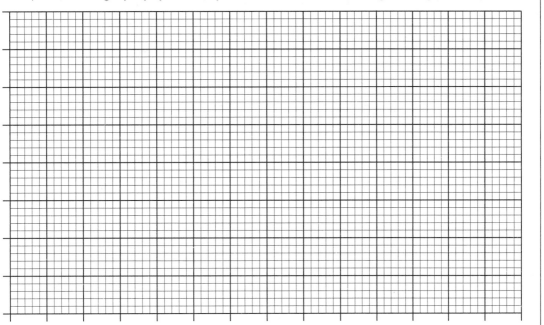

(d) Give a conclusion about the suitability of the donors.

1

_____

(e) (i) The base sequence of a primer used in the PCR procedure is shown below.

A T G A C A A A T C G

Give the base sequence of a DNA fragment to which this primer would bind.

1

_____

(ii) Complete the table below to show the temperatures used in two stages of the PCR procedure and the reasons for using these temperatures.

2

| Temperature (°C) | Reason |
|---|---|
| 94 | |
| | Allows primer to bind to target sequence |

MARKS | DO NOT WRITE IN THIS MARGIN

12. An investigation was carried out involving a number of patients with heart disease. A group of volunteer patients was treated with adult stem cells and a control group was not given this treatment.

Six weeks after the treatment, the average heart rate and the average volume of blood pumped out per heartbeat (stroke volume) was determined for each group.

The results are shown in the table below.

| | Patients given stem cell treatment | Patients not given stem cell treatment |
|---|---|---|
| Average heart rate (beats per minute) | 70 | 70 |
| Average stroke volume (cm$^3$) | 45 | 28 |

(a) Give **two** conclusions which can be drawn about the effect of the stem cell treatment on the patients.

2

1 _____

_____

2 _____

_____

(b) Another important measure of heart performance is cardiac output.

Cardiac output = heart rate × stroke volume

(cm$^3$ per minute)    (bpm)        (cm$^3$)

Calculate the average increase in cardiac output in those patients given the stem cell treatment compared to those in the control group.

1

*Space for calculation*

_____ cm$^3$ per minute

MARKS | DO NOT WRITE IN THIS MARGIN

12. (continued)

(c) (i) Describe how tissue (adult) stem cells differ from embryonic stem cells.

1

_____

_____

(ii) Describe how the heart cells produced by the patients as a result of the stem cell treatment in this investigation developed their specialised functions.

1

_____

_____

(d) Much stem cell research is related to the therapeutic value of stem cells.

Give **one other** reason for carrying out stem cell research.

1

_____

_____

[Turn over for Question 13 on *Page thirty*

MARKS | DO NOT WRITE IN THIS MARGIN

13. Answer **either A or B** in the space below and on pages *thirty-one* and *thirty-two*.

   **Labelled diagrams may be used where appropriate.**

   A  Write notes on gene expression in eukaryotes under the following headings:

       (i)   production of mRNA;    **5**

       (ii)  translation of mRNA.    **4**

   **OR**

   B  Write notes on mutation under the following headings:

       (i)   single gene mutations;    **4**

       (ii)  chromosome mutations and polyploidy.    **5**

**SPACE FOR ANSWERS**

SPACE FOR ANSWERS

[END OF QUESTION PAPER]

**ADDITIONAL SPACE FOR ANSWERS AND ROUGH WORK**

ADDITIONAL GRAPH PAPER FOR QUESTION 11(c)

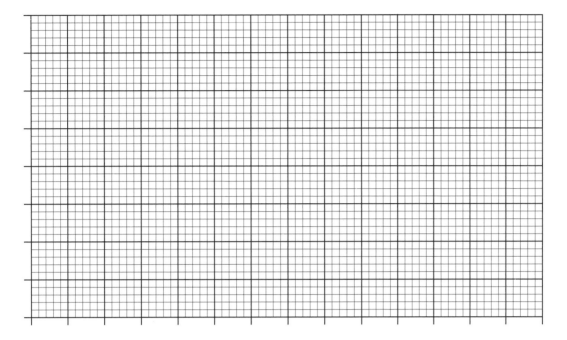

**ADDITIONAL SPACE FOR ANSWERS AND ROUGH WORK**

HIGHER

2016

National
Qualifications
2016

**X707/76/02**

**Biology**
**Section 1 — Questions**

MONDAY, 9 MAY
9:00 AM — 11:30 AM

Instructions for the completion of Section 1 are given on *Page two* of your question and answer booklet X707/76/01.

Record your answers on the answer grid on *Page three* of your question and answer booklet.

Before leaving the examination room you must give your question and answer booklet to the Invigilator; if you do not, you may lose all the marks for this paper.

**SECTION 1 — 20 marks**

**Attempt ALL questions**

1. The diagram below shows part of a DNA molecule before and after a mutation.

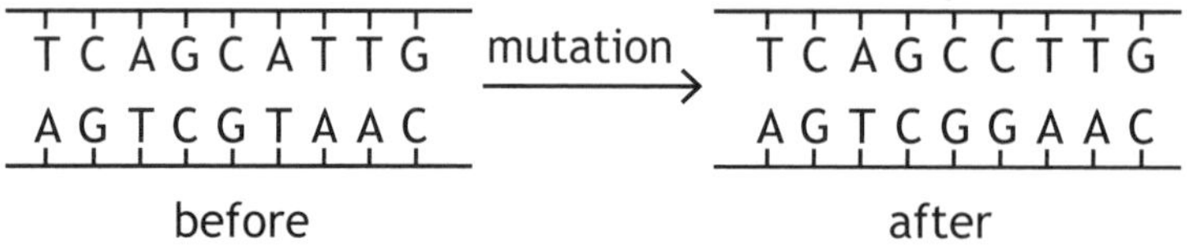

TCAGCATTG　mutation　TCAGCCTTG
AGTCGTAAC　───→　AGTCGGAAC
　　before　　　　　after

The type of mutation shown is

A   deletion

B   substitution

C   insertion

D   inversion.

2. Which of the following are required in a polymerase chain reaction (PCR)?

A   DNA polymerase, template strand and primers

B   RNA polymerase, template strand and primers

C   DNA polymerase, template strand and ligase

D   RNA polymerase, ligase and primers

3. Each cycle of a polymerase chain reaction (PCR) takes 5 minutes.

If there are 1000 DNA fragments at the start of the reaction, how long will it take for the number of fragments produced by the reaction to be greater than 1 million?

A   15 minutes

B   35 minutes

C   50 minutes

D   55 minutes

4. The graphs below show possible changes in the body size of a population of barn swallows, *Hirudino rusticana*, in response to a selection pressure.

————— original population

············ population after selection

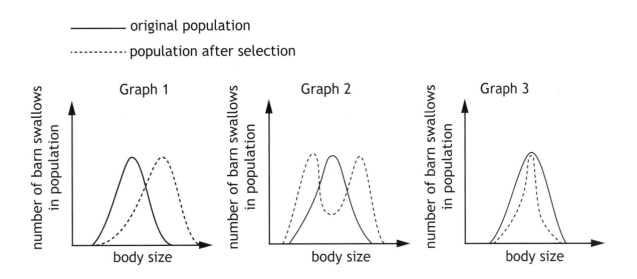

Which row in the table below matches each graph with the type of selection taking place?

| | | Graph | |
|---|---|---|---|
| | *1* | *2* | *3* |
| A | disruptive | directional | stabilising |
| B | directional | disruptive | stabilising |
| C | stabilising | disruptive | directional |
| D | directional | stabilising | disruptive |

[Turn over

5.  The diagram below represents a phylogenetic tree showing the evolutionary relatedness of several species of cat.

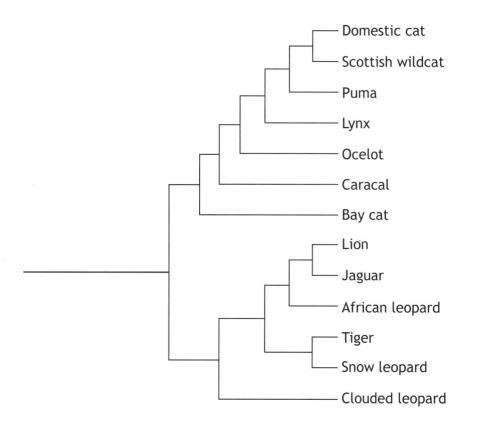

With how many species does the African leopard share a common ancestor in this phylogenetic tree?

A    2 only

B    5 only

C    12 only

D    13

6.  Over millions of years of evolution, mutations occur at a broadly constant rate within a gene. This allows genes to be used as molecular clocks. The diagram below shows how the base sequence in part of a gene changed as two evolutionary lineages diverged from an original base sequence. The base sequence in the gene has changed at a rate of 1 base per 5 million years as shown.

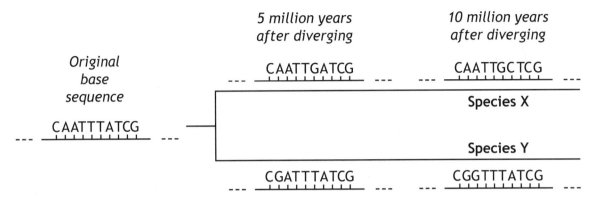

    Assuming this rate of mutation continued, by how many bases would this part of the gene differ in Species X compared with Species Y 20 million years after diverging from the original base sequence?

    A    4

    B    8

    C    16

    D    20

7.  In metabolic pathways, the rates of reaction can be affected by the presence of enzyme inhibitors.

    Which row in the table below is correct?

| | Type of inhibition | Inhibitor binds to active site | Effect of increasing substrate concentration on inhibition |
|---|---|---|---|
| A | competitive | yes | reversed |
| B | non competitive | yes | unaffected |
| C | competitive | no | unaffected |
| D | non competitive | no | reversed |

8.  Which row in the table below identifies the number of heart chambers and the type of circulatory system in amphibians?

| | Number of heart chambers | Type of circulatory system |
|---|---|---|
| A | 3 | incomplete double |
| B | 4 | incomplete double |
| C | 3 | complete double |
| D | 4 | complete double |

[Turn over

9. During unexpected periods of drought the South American lungfish, *Lepidosiren paradoxa*, survives by burying into mud.

   This type of behaviour is known as

   A    predictive dormancy

   B    daily torpor

   C    aestivation

   D    hibernation.

10. An experiment was set up to investigate the effect of different respiratory substrates on the rate of respiration in yeast. Methylene blue can be used to measure the rate of respiration as it changes from dark blue to colourless when it accepts hydrogen ions. Four test tubes were set up, each containing yeast, methylene blue and one of the respiratory substrates.

    The table below shows the results of this investigation.

| Test tube number | Respiratory substrate | Appearance of the methylene blue after 20 minutes |
|---|---|---|
| 1 | starch | dark blue |
| 2 | sucrose | light blue |
| 3 | lactose | dark blue |
| 4 | glucose | colourless |

   Which of the following conclusions is correct?

   The rate of respiration is

   A    higher with starch than with glucose

   B    lower with sucrose than with lactose

   C    higher with glucose than with lactose

   D    lower with glucose than with sucrose.

11. Stages of aerobic respiration are shown below.

    1    Glycolysis
    2    Citric acid cycle
    3    Electron transfer chain

    Which stage(s) involve(s) **both** phosphorylation of intermediates and generation of ATP?

    A    1 only

    B    3 only

    C    1 and 2 only

    D    1 and 3 only

12.  Which row in the table below identifies a stage of aerobic respiration, its site and an event which occurs during that stage?

|   | Stage | Site | Event |
|---|-------|------|-------|
| A | electron transfer chain | inner mitochondrial membrane | carbon dioxide is released |
| B | electron transfer chain | matrix of mitochondrion | hydrogen ions combine with oxygen |
| C | citric acid cycle | inner mitochondrial membrane | hydrogen ions combine with oxygen |
| D | citric acid cycle | matrix of mitochondrion | carbon dioxide is released |

13.  A field trial was set up to investigate the effect of phosphate fertiliser on the yield of the potato cultivar Maris Piper. Potatoes were planted in 5 plots, each of which received a different level of phosphate fertiliser. When they were harvested the yield from each plot was recorded.

A list of suggested improvements to this field trial is shown below.

1    Apply equal volumes of water to each plot.
2    Grow the same number of potato plants in each plot.
3    Use 10 plots at each phosphate fertiliser level.
4    Plant different potato cultivars in each plot.

Which of the suggestions would improve the validity of the results?

A    1 and 2

B    1 and 3

C    2 and 4

D    3 and 4

[Turn over

14. Which compound combines with hydrogen during carbon fixation (Calvin cycle)?

   A   Ribulose biphosphate

   B   NADP

   C   Oxygen

   D   3-phosphoglycerate

15. The following absorption spectra were obtained by testing four different plant extracts.

   Which extract contains chlorophyll?

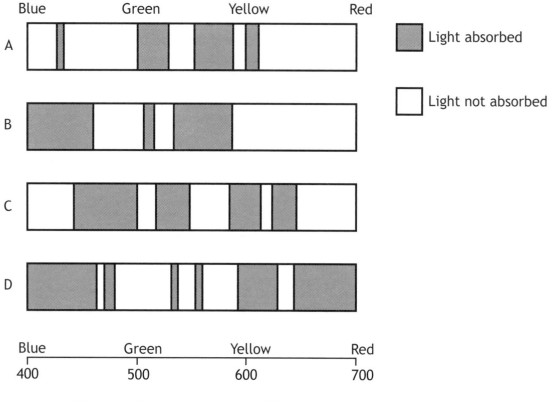

Wavelength (nm) and colour of light

**16.** The table below shows the biological and economic yields of four different crops.

| Crop | Biological yield (tonnes of dry mass/hectare) | Economic yield (tonnes of dry mass/hectare) |
|---|---|---|
| pea | 10 | 2 |
| rice | 15 | 10 |
| wheat | 30 | 8 |
| potato | 30 | 10 |

The crop with the highest harvest index is

A    pea

B    rice

C    wheat

D    potato.

**17.** The graph below shows the levels of nitrogen and phosphorus applied to crops in an area of Scotland between 1986 and 2006.

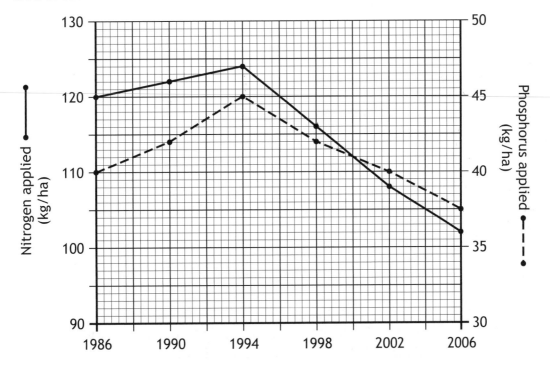

In which year was there the smallest difference between the levels of nitrogen and phosphorus applied?

A    1998

B    2000

C    2002

D    2006

[Turn over

18.  Which of the following are features of naturally inbreeding crop plants?

1    Susceptible to inbreeding depression
2    Deleterious alleles eliminated by natural selection
3    Self-pollinating

A    1 and 2 only

B    1 and 3 only

C    2 and 3 only

D    1, 2 and 3

19.  On returning to their roost after feeding, vampire bats may regurgitate blood to feed an unrelated individual in the same social group.

This is an example of

A    mutualism

B    altruism

C    social hierarchy

D    kin selection.

20.  The statements below refer to behaviour sometimes displayed by lions kept in captivity.

1    Repetitive chewing on cage bars
2    Excessive licking of body
3    Continually pacing backwards and forward

Which are examples of stereotypy?

A    1 only

B    1 and 2 only

C    2 and 3 only

D    1, 2 and 3

**[END OF SECTION 1.  NOW ATTEMPT THE QUESTIONS IN SECTION 2
OF YOUR QUESTION AND ANSWER BOOKLET]**

# H

## National Qualifications 2016

Mark

**X707/76/01**

# Biology
## Section 1 — Answer Grid and Section 2

MONDAY, 9 MAY

9:00 AM — 11:30 AM

**Fill in these boxes and read what is printed below.**

Full name of centre

Town

Forename(s)

Surname

Number of seat

Date of birth

| Day | Month | Year | Scottish candidate number |
|-----|-------|------|---------------------------|

**Total marks — 100**

**SECTION 1 — 20 marks**

Attempt ALL questions.

Instructions for the completion of Section 1 are given on *Page two*.

**SECTION 2 — 80 marks**

Attempt ALL questions.

Questions 7 and 14 contain a choice.

Write your answers clearly in the spaces provided in this booklet. Additional space for answers and rough work is provided at the end of this booklet. If you use this space you must clearly identify the question number you are attempting. Any rough work must be written in this booklet. You should score through your rough work when you have written your final copy.

Use **blue** or **black** ink.

Before leaving the examination room you must give this booklet to the Invigilator; if you do not, you may lose all the marks for this paper.

## SECTION 1 — 20 marks

The questions for Section 1 are contained in the question paper X707/76/02.

Read these and record your answers on the answer grid on *Page three* opposite.

Use **blue** or **black** ink. Do NOT use gel pens or pencil.

1. The answer to each question is **either** A, B, C or D. Decide what your answer is, then fill in the appropriate bubble (see sample question below).

2. There is **only one correct** answer to each question.

3. Any rough working should be done on the additional space for answers and rough work at the end of this booklet.

## Sample Question

The thigh bone is called the

     A    humerus

     B    femur

     C    tibia

     D    fibula.

The correct answer is **B** — femur. The answer **B** bubble has been clearly filled in (see below).

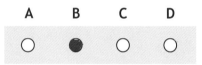

## Changing an answer

If you decide to change your answer, cancel your first answer by putting a cross through it (see below) and fill in the answer you want. The answer below has been changed to **D**.

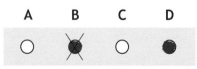

If you then decide to change back to an answer you have already scored out, put a tick (✓) to the **right** of the answer you want, as shown below:

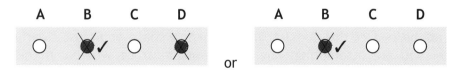

or

## SECTION 1 — Answer Grid

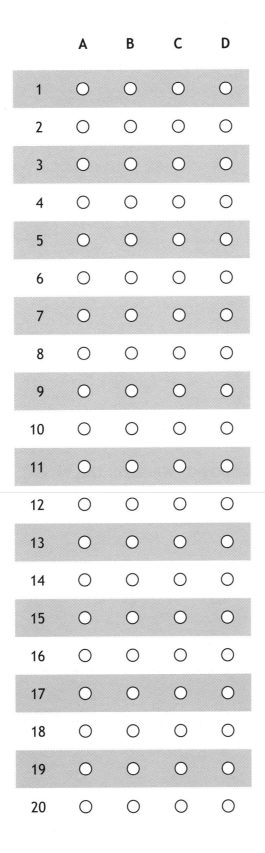

[BLANK PAGE]

DO NOT WRITE ON THIS PAGE

[Turn over for next question

**DO NOT WRITE ON THIS PAGE**

SECTION 2 — 80 marks

**Attempt ALL questions**

Questions 7 and 14 each contain a choice.

MARKS | DO NOT WRITE IN THIS MARGIN

1. The diagram below shows a process involved in the production of a polypeptide in a cell.

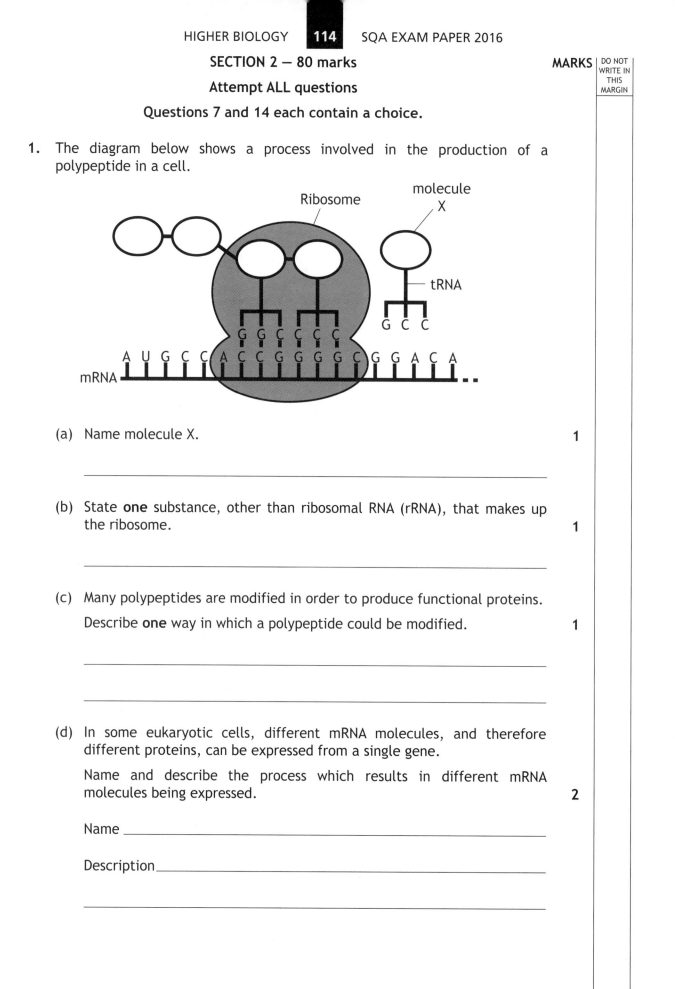

(a) Name molecule X.

1

_____

(b) State **one** substance, other than ribosomal RNA (rRNA), that makes up the ribosome.

1

_____

(c) Many polypeptides are modified in order to produce functional proteins.

Describe **one** way in which a polypeptide could be modified.

1

_____

_____

(d) In some eukaryotic cells, different mRNA molecules, and therefore different proteins, can be expressed from a single gene.

Name and describe the process which results in different mRNA molecules being expressed.

2

Name _____

Description_____

_____

MARKS | DO NOT WRITE IN THIS MARGIN

**2.** DNA holds the genetic information in both prokaryotic and eukaryotic cells.

   (a)   (i)   Describe **one** organisational difference between prokaryotic and eukaryotic chromosomal DNA.

1

          (ii)   Name the substance with which DNA is packaged in eukaryotes.

1

   (b)   State **one** location, other than the nucleus, where DNA is found in eukaryotic cells.

1

   (c)   During DNA replication two new daughter strands are synthesised using the original strands as templates.

          (i)   State why the antiparallel nature of the DNA molecule results in one of the strands being synthesised in short fragments.

1

          (ii)   Template DNA, enzymes and ATP are necessary for DNA replication.

               State **one** other substance required.

1

   (d)   Explain why cells need to carry out DNA replication.

1

[Turn over

MARKS | DO NOT WRITE IN THIS MARGIN

3. Stem cells are unspecialised cells which can be found in embryonic and adult tissue.

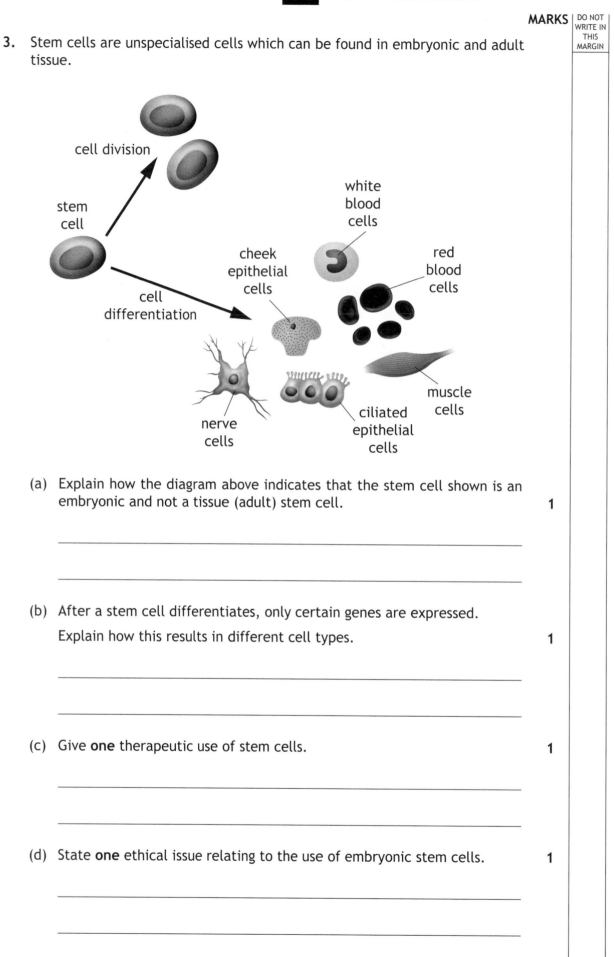

(a) Explain how the diagram above indicates that the stem cell shown is an embryonic and not a tissue (adult) stem cell.

1

_____

_____

(b) After a stem cell differentiates, only certain genes are expressed.

Explain how this results in different cell types.

1

_____

_____

(c) Give **one** therapeutic use of stem cells.

1

_____

_____

(d) State **one** ethical issue relating to the use of embryonic stem cells.

1

_____

_____

[Turn over for next question

**DO NOT WRITE ON THIS PAGE**

MARKS | DO NOT WRITE IN THIS MARGIN

4. Meristems can be cultured in growth medium to produce new plants.

An experiment was carried out to investigate the effects of three different growth media (A, B and C) on the production of shoots by meristems of African violet plants.

Five meristems were removed and cultured in each medium for a period of seven weeks. The average number of shoots produced per meristem was recorded at specific times during the investigation.

The results are shown in the graph below.

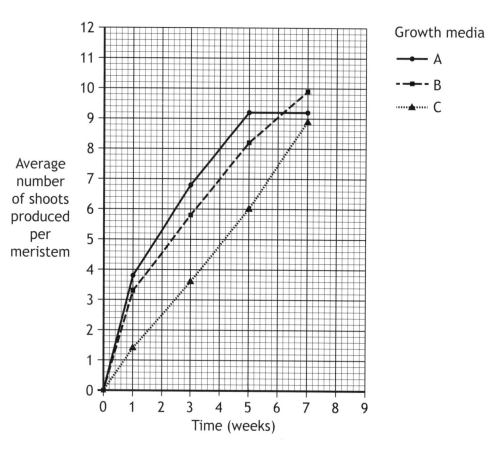

(a)  (i)  **Use values from the graph** to describe the average number of shoots produced per meristem over the seven week period in medium A.

2

_____

_____

_____

(ii) Calculate the percentage increase in the average number of shoots produced per meristem between week 1 and week 7 in medium B.

1

*Space for calculation*

_____ %

MARKS | DO NOT WRITE IN THIS MARGIN

4.  (a) (continued)

(iii) Table 1 below shows the number of shoots produced per meristem at three weeks in one of the media.

**Table 1**

| Meristem | Number of shoots produced per meristem |
|----------|----------------------------------------|
| 1 | 4 |
| 2 | 5 |
| 3 | 7 |
| 4 | 7 |
| 5 | 6 |

Using information from **Table 1 and the graph**, state the medium in which these meristems were cultured.

1

*Space for calculation*

Medium_____

(b) Predict which medium would produce plants with the greatest number of shoots after nine weeks growth. Give a reason for your answer.

2

Medium _____

Reason _____

_____

(c) In a further experiment, the average number of roots and average root length at 7 weeks were recorded in each of the media.

The results are shown in Table 2 below.

**Table 2**

| Medium | Average number of roots produced per meristem | Average root length (mm) |
|--------|-----------------------------------------------|--------------------------|
| A | 12 | 12 |
| B | 11 | 19 |
| C | 12 | 17 |

After analysing the results, medium B was used for the commercial production of plants.

**Use the information in Table 2** to explain why plants cultured in medium B would grow best.

2

_____

_____

MARKS | DO NOT WRITE IN THIS MARGIN

5.  In the North Pacific Ocean there are two different populations of killer whales *Orcinus orca*. One population feeds mainly on fish while the other feeds mainly on sea mammals.

This behavioural barrier has led to considerable genetic variation between these populations.

(a) (i) Name the type of speciation which could occur as a result of this barrier.    **1**

_____

(ii) State the importance of isolation barriers in speciation.    **1**

_____

_____

(iii) Scientists believe that these two populations are still the same species.

Suggest how they could confirm this.    **1**

_____

_____

(b) Polyploidy can lead to speciation.

(i) State what is meant by the term polyploidy.    **1**

_____

_____

(ii) Describe **one** example of the importance of polyploidy in evolution.    **1**

_____

_____

MARKS | DO NOT WRITE IN THIS MARGIN

6.  The antibiotic bacitracin is produced by the bacterial species *B.subtilis*.

The graph below shows the growth curve of a population of *B.subtilis* cultured to produce the antibiotic.

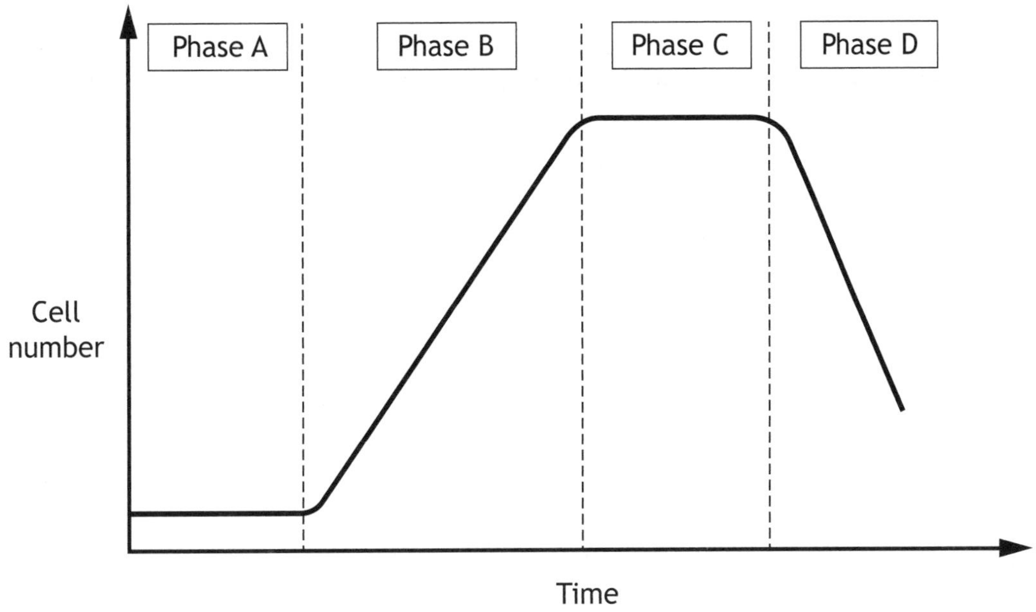

(a)  Name Phase A and explain why cells do not divide during this phase.        2

Name _____

Explanation _____

_____

(b)  (i)  Name the phase in which the bacteria produce the secondary metabolite bacitracin.        1

_____

(ii)  Explain why this secondary metabolite gives an ecological advantage to *B.subtilis*.        1

_____

_____

(c)  This growth curve shows viable cell numbers of *B.subtilis*.

Give evidence from the graph to justify this statement.        1

_____

_____

MARKS | DO NOT WRITE IN THIS MARGIN

7. Answer **either A or B** in the space below.

   A   Describe and compare anabolic and catabolic reactions.   4

   **OR**

   B   Describe and compare metabolism in conformers and regulators.   4

[Turn over for next question

**DO NOT WRITE ON THIS PAGE**

8. An investigation was set up to monitor growth of bacteria in compost. The compost was added to a fermenter and the temperature of the compost was recorded over a 20 day period. Samples of the compost were cultured and the numbers of three bacterial species present were recorded.

The compost temperatures and the populations of the three species of bacteria are shown in the table below.

| Time (days) | Compost Temperature (°C) | Population (millions per gram of compost) | | |
|---|---|---|---|---|
| | | Species A | Species B | Species C |
| 0 | 21 | 396·0 | 0·4 | 123·0 |
| 2 | 40 | 4·2 | 10·2 | 14·6 |
| 4 | 72 | 0·1 | 195·0 | 0·1 |
| 6 | 53 | 0 | 8·5 | 0 |
| 20 | 32 | 0 | 0 | 0 |

(a) Calculate how many times greater the population of Species A was compared to Species B at the start of the investigation.

*Space for calculation*

1

(b) Describe the relationship between temperature of the compost and population of Species C over the first four days.

1

MARKS | DO NOT WRITE IN THIS MARGIN

8.  (continued)

(c)    (i)  **Using information in the table**, state which species of bacteria is thermophilic and justify your answer.    2

Species _____

Justification _____

_____

(ii)  Describe how thermophilic bacteria are adapted to survive in their environment.    1

_____

_____

(iii)  Give an example, other than in compost, of an environment where thermophilic bacteria are adapted to grow successfully.    1

_____

[Turn over

MARKS | DO NOT WRITE IN THIS MARGIN

9. The diagram below shows how a human gene can be inserted into bacteria to produce human insulin using recombinant DNA technology.

(a) Name **one** enzyme used in this process and state its function.    2

Name _____

Function _____

(b) (i) The recombinant plasmid also contains a gene for resistance to the antibiotic, ampicillin.

Describe a procedure which would allow only cells containing the recombinant plasmid to be selected.    2

_____

_____

_____

(ii) Plasmids with these antibiotic resistance genes have been passed to other bacterial species by horizontal transfer.

Describe the process of horizontal transfer.    1

_____

_____

(c) When culturing the bacteria which produce insulin, sterile conditions are maintained.

Explain why this is important.    1

_____

_____

[Turn over for next question

DO NOT WRITE ON THIS PAGE

MARKS | DO NOT WRITE IN THIS MARGIN

10. An investigation was carried out to compare the rate of metabolism in a species of cricket, *Gryllus assimilis*, at different temperatures.

Five crickets were placed in a sealed flask which was fitted with a carbon dioxide ($CO_2$) sensor as shown in the diagram below.

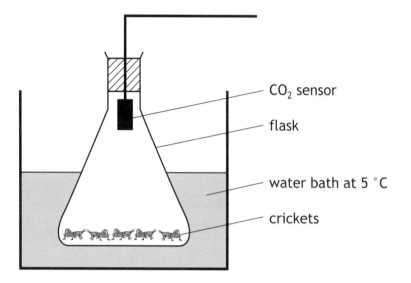

$CO_2$ sensor

flask

water bath at 5 °C

crickets

The flask was placed in a water bath at 5 °C and left for 10 minutes.

The $CO_2$ produced per minute was then measured. This procedure was repeated at 10, 15, 20 and 30 °C.

The results are shown in the table below.

| Temperature (°C) | Rate of $CO_2$ production (units per minute) |
|---|---|
| 5 | 300 |
| 10 | 500 |
| 15 | 800 |
| 20 | 1200 |
| 30 | 1600 |

(a) (i) Give a reason why the flask was left for 10 minutes at each temperature **before** each reading was taken.          1

_____

_____

MARKS | DO NOT WRITE IN THIS MARGIN

**10.**   **(a)  (continued)**

(ii)  A control flask should be included in this investigation.

Describe the control and explain its purpose in the investigation.    **2**

Description _____

_____

Explanation _____

_____

(b)  Plot a line graph to show the results of the investigation.    **2**

(Additional graph paper, if required can be found on *Page thirty-one*).

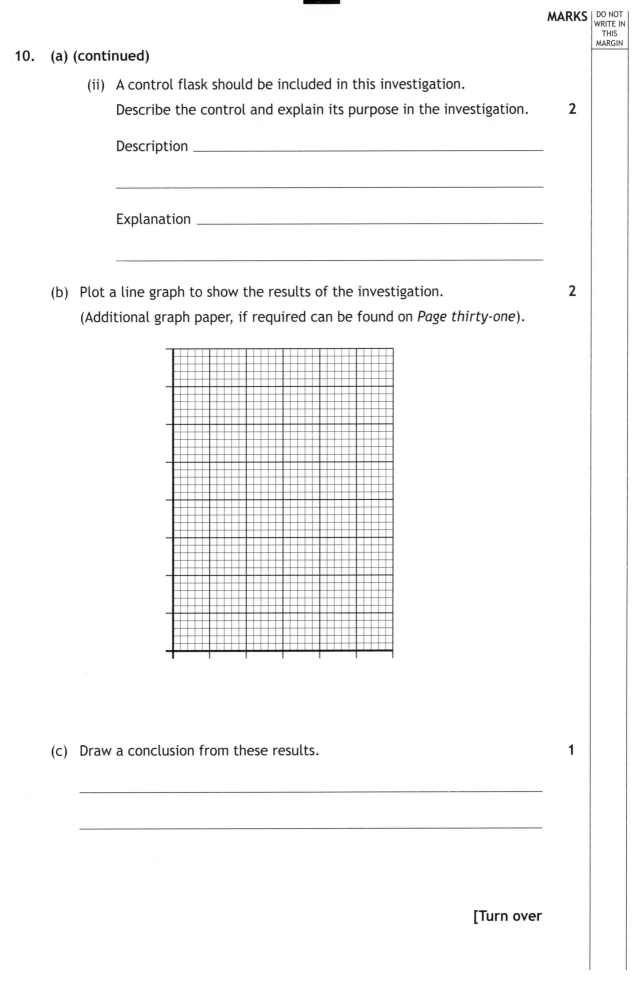

(c)  Draw a conclusion from these results.    **1**

_____

_____

**[Turn over**

MARKS | DO NOT WRITE IN THIS MARGIN

11. Colchicine is a chemical used in plant breeding programmes to induce mutations and produce cultivars with improved characteristics.

Sesame is an important crop plant grown for its edible seeds and leaves.

An investigation was carried out to determine the effects of colchicine concentration on sesame. Sesame seeds were soaked in different concentrations of colchicine solution for 24 hours. Seeds from each concentration were germinated and 50 plants were grown from each concentration. Ninety days later the total leaf area, number of seeds and mass of seeds per plant were recorded.

The average results are shown in the table below.

| Colchicine concentration $(m \, mol \, l^{-1})$ | Average total leaf area per plant $(cm^2)$ | Average number of seeds per plant | Average total mass of seeds per plant $(g)$ |
|---|---|---|---|
| 0 | 1500 | 548 | 2·8 |
| 0·1 | 2315 | 532 | 3·5 |
| 0·5 | 2786 | 550 | 4·4 |
| 1·0 | 3500 | 512 | 4·7 |

(a) (i) Identify the independent variable in this investigation. 1

_____

(ii) State an aspect of the investigation which helped to ensure that reliable results were obtained. 1

_____

(b) (i) An important characteristic of food crops is the *1000 seed mass* which is the total mass of a sample of 1000 seeds.

Calculate the *1000 seed mass* for the plants grown from seeds soaked in a colchicine concentration of 0·5 m mol $l^{-1}$. 1

*Space for calculation*

_____ g

MARKS | DO NOT WRITE IN THIS MARGIN

11. **(b) (continued)**

(ii) Express, as the simplest whole number ratio, the average total leaf area per plant from seeds soaked in a colchicine concentration of 0 to that at $1\cdot0$ m mol $l^{-1}$.

1

*Space for calculation*

_____ : _____

0                                   $1\cdot0$
m mol $l^{-1}$                      m mol $l^{-1}$

(c) Explain the relationship between the total leaf area and total mass of seeds.

2

_____

_____

_____

[Turn over

MARKS | DO NOT WRITE IN THIS MARGIN

12. Malaria is a disease in humans caused by a parasite which is transmitted from human to human by mosquitoes. The stages of infection in humans are shown in the flow diagram below.

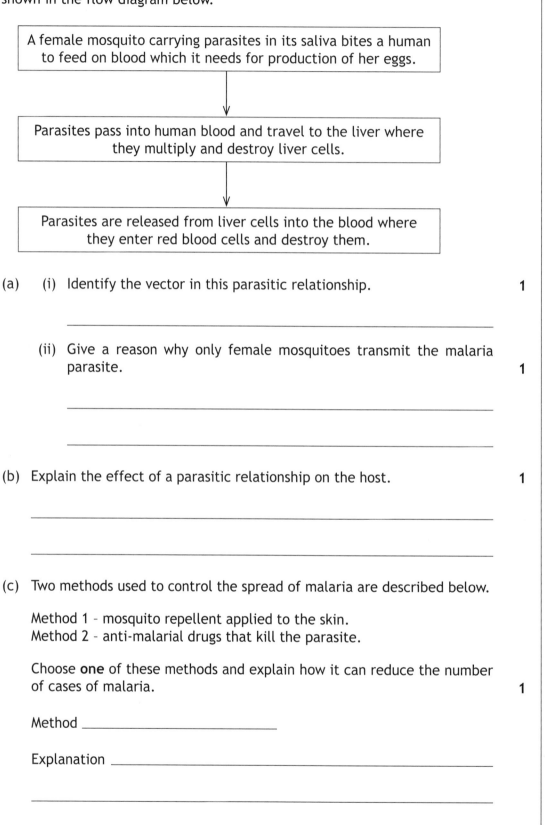

A female mosquito carrying parasites in its saliva bites a human to feed on blood which it needs for production of her eggs.

↓

Parasites pass into human blood and travel to the liver where they multiply and destroy liver cells.

↓

Parasites are released from liver cells into the blood where they enter red blood cells and destroy them.

(a) (i) Identify the vector in this parasitic relationship.

1

_____

(ii) Give a reason why only female mosquitoes transmit the malaria parasite.

1

_____

_____

(b) Explain the effect of a parasitic relationship on the host.

1

_____

_____

(c) Two methods used to control the spread of malaria are described below.

Method 1 – mosquito repellent applied to the skin.
Method 2 – anti-malarial drugs that kill the parasite.

Choose **one** of these methods and explain how it can reduce the number of cases of malaria.

1

Method _____

Explanation _____

_____

[Turn over for next question

**DO NOT WRITE ON THIS PAGE**

13. Freshwater mussels are small animals which live on the beds of lakes and rivers. Zebra mussels are a species of freshwater mussel native to lakes in Russia. They were accidentally introduced by humans into a river in North America in 1991. The populations of zebra mussels and the native unionid mussels were measured over a 12 year period.

The results are shown in the graph below.

(a)  (i)  State the unionid mussel population in 1993.                                      1

_____ mussels per m$^2$

(ii)  State the zebra mussel population when the unionid mussel population was 50 mussels per m$^2$.                                      1

_____ mussels per m$^2$

(iii)  Calculate the average increase per year in the zebra mussel population between 1991 and 2003.                                      1

*Space for calculation*

_____ mussels per m$^2$ per year

(b)  Explain how the graph confirms that zebra mussels are more successful competitors than unionid mussels.                                      1

_____

_____

MARKS | DO NOT WRITE IN THIS MARGIN

13. **(continued)**

(c) **Using evidence from the graph**, explain why zebra mussels are an invasive species.

1

_____

_____

(d) Suggest a reason why the population of zebra mussels may have increased faster in the North American river than in its native habitat.

1

_____

_____

(e) Invasive species have a negative impact on genetic diversity of an ecosystem.

State what is meant by genetic diversity.

1

_____

_____

[Turn over for next question

MARKS | DO NOT WRITE IN THIS MARGIN

14. Answer **either A or B** in the space below and on *Pages 29* and *30*.

   A   Write notes on crop protection under the following headings:

   (i)   weeds, pests and diseases;                                                4

   (ii)  methods of control.                                                        4

   **OR**

   B   Write notes on social behaviour in animals under the following headings:

   (i)   social hierarchy and cooperative hunting;                                 4

   (ii)  social insects.                                                           4

**SPACE FOR ANSWERS**

**SPACE FOR ANSWERS**

[END OF QUESTION PAPER]

**ADDITIONAL SPACE FOR ANSWERS AND ROUGH WORK**

ADDITIONAL GRAPH PAPER FOR QUESTION 10 (b)

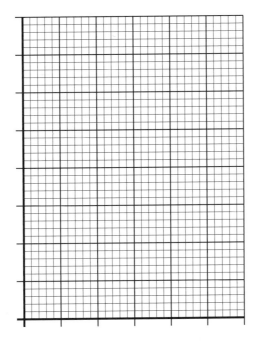

MARKS | DO NOT WRITE IN THIS MARGIN

## ADDITIONAL SPACE FOR ANSWERS AND ROUGH WORK

# SQA HIGHER BIOLOGY 2016

## HIGHER BIOLOGY
## 2014 SPECIMEN QUESTION PAPER

### Section 1

| Question | Response | Mark |
|---|---|---|
| 1. | C | 1 |
| 2. | B | 1 |
| 3. | C | 1 |
| 4. | A | 1 |
| 5. | D | 1 |
| 6. | A | 1 |
| 7. | A | 1 |
| 8. | C | 1 |
| 9. | B | 1 |
| 10. | D | 1 |
| 11. | A | 1 |
| 12. | A | 1 |
| 13. | B | 1 |
| 14. | A | 1 |
| 15. | D | 1 |
| 16. | C | 1 |
| 17. | D | 1 |
| 18. | D | 1 |
| 19. | A | 1 |
| 20. | B | 1 |

### Section 2

| Question | | | Expected response | Max mark |
|---|---|---|---|---|
| 1. | (a) | (i) | Intron/Intron 1/Intron 2 | 1 |
| | | (ii) | (Alternative) RNA splicing | 1 |
| | | (iii) | • Depending on which RNA segments are treated as exons and introns  1 different segments can be spliced together to produce different mRNA transcripts  1 OR • Appropriate example from diagram | 2 |
| | (b) | | • Cutting and combining different protein chains OR • Adding phosphate to the protein OR • Adding carbohydrate to the protein | 1 |
| 2. | (a) | | Translocation | 1 |
| | (b) | (i) | Competitive | 1 |
| | | (ii) | 95 | 1 |

| Question | | | Expected response | Max mark |
|---|---|---|---|---|
| | | (iii) | • Drug was effective as white blood count reduced to normal • Drug works by inhibiting the enzyme produced by Philadelphia chromosome | 2 |
| 3. | (a) | | • Stage 1 separates strands or breaks H bonds • Stage 2 allows primer to bond/anneal to strand/target sequence | 2 |
| | (b) | | 7 | 1 |
| | (c) | | Identical set up but without primers | 1 |
| | (d) | | Forensic use/paternity testing | 1 |
| 4. | (a) | | Sequence data | 1 |
| | (b) | | Horizontal/lateral | 1 |
| | (c) | (i) | 25 | 1 |
| | | (ii) | • Last common ancestor of humans and chimpanzees was more recent than humans and orangutans • Chimpanzees and humans 5 million years ago, orangutans and humans 19 million years ago | 2 |
| 5. | (a) | | • P is Acetyl CoA • Q is Oxaloacetate | 2 |
| | (b) | | • ATP/Energy is required  1 • A greater amount of energy/ATP is produced  1 | 2 |
| | (c) | | • Carry hydrogen and high energy electrons  1 • To the electron transport chain  1 | 2 |
| | (d) | | • Less ATP/energy is produced  1 • Fewer electrons are passed to electron transport chain OR • Fewer hydrogen ions are pumped through the membrane OR • ATP synthase is damaged  1 | 2 |
| 6. | (a) | | 20 | 1 |
| | (b) | | • Increase - people becoming complacent about hand washing or bacteria becoming resistant OR • No change - everyone now using procedure OR • Decrease - increased uptake of procedure | 1 |
| | (c) | | • *Clostridium* increases • *Staphylococcus* remains fairly constant | 2 |
| | (d) | | • Conclusion - effective • Justification - although percentage of cases remains similar number of cases falls | 2 |

| Question | | | Expected response | Max mark |
|---|---|---|---|---|
| | (e) | | • Type - *Clostridium*<br>• Reason - percentage of cases due to *Clostridium* increased | 1 |
| 7. | (a) | | Enzymes have an optimum temperature or only work within a certain temperature range | 1 |
| | (b) | (i) | Hypothalamus | 1 |
| | | (ii) | Nerve (impulse) | 1 |
| | (c) | (i) | Vasoconstriction/vessels get narrower | 1 |
| | | (ii) | Reduces blood flow to skin so less heat loss | 1 |
| 8. | A | | 1. metabolic rate reduced<br>2. dormancy can be predictive or consequential<br>3. hibernation in winter usually mammals<br>4. aestivation allows survival in periods of drought or high temperature<br>5. daily torpor is reduced activity in animals with high metabolic rates<br>6. example of hibernation or aestivation or daily torpor | 4 |
| | B | | 1. plant/animal gene transferred into microorganism that makes plant/animal protein<br>2. restriction endonuclease to cut gene out/cut plasmid<br>3. genes introduced to increase yield or prevent microbe surviving in external environment<br>4. ligase seals gene into plasmid<br>5. recombinant yeast cells to overcome polypeptides being incorrectly folded or lacking post translational modifications<br>6. regulatory sequences in plasmids/artificial chromosomes to control gene expression | 4 |
| 9. | (a) | | 15 | 1 |
| | (b) | | 413·44 | 1 |
| | (c) | | • Milk yield/fat content increased by crossbreeding<br>• Protein content decreased by crossbreeding | 1 |
| | (d) | | Inbreeding depression | 1 |
| | (e) | (i) | F2 has a variety of genotypes | 1 |
| | | (ii) | Selection or backcrossing | 1 |
| 10. | (a) | | Rate of photosynthesis | 1 |
| | (b) | | Use a water bath | 1 |
| | (c) | | • Easier to separate algae from solution<br>OR<br>• Easier to control algae concentration | 1 |
| | (d) | | Repeat at each distance/light intensity | 1 |
| | (e) | | • Axes and labels<br>• Plotting and joined with a ruler | 2 |

| Question | | | Expected response | Max mark |
|---|---|---|---|---|
| | (f) | | • As light intensity increases rate increases<br>• At higher light intensities rate remains constant | 2 |
| 11. | (a) | | • Worker bees are related to queen's offspring<br>• So worker bees share genes with queen's offspring | 2 |
| | (b) | (i) | Increase from 4.2 million (in 1980) to 4.4 million (in 1985) then decrease to 2.8 million (in 1995) | 1 |
| | | (ii) | 2:1 | 1 |
| 12. | (a) | (i) | Number/frequency of alleles in a population | 1 |
| | | (ii) | • Small population may lose the genetic variation necessary to enable evolutionary responses to environmental change<br>OR<br>• The loss of genetic diversity can lead to inbreeding which results in poor reproductive rates | 1 |
| | (b) | | Edge species may invase the interior of the habitat and compete with interior species | 1 |
| | (c) | (i) | Area of natural habitat linking fragments | 1 |
| | | (ii) | Individual members of the locally extinct species can move into the fragment and recolonise | 1 |
| 13. | (a) | | Invasive | 1 |
| | (b) | | • Light<br>OR<br>• Water<br>OR<br>• Minerals<br>OR<br>• Nutrients | 1 |
| | (c) | (i) | • May eat native plants<br>OR<br>• May become invasive | 1 |
| | | (ii) | Test effect on native species in an enclosed area | 1 |

| Question | | | Expected response | Max mark |
|---|---|---|---|---|
| 14. | A | | 1. double strand of nucleotides/ double helix<br>2. deoxyribose sugar, phosphate and base<br>3. sugar phosphate backbone<br>4. complementary bases pair or A-T and C-G<br>5. H bonds between bases<br>6. antiparallel structure with deoxyribose and phosphate at 3' and 5' ends<br>7. proteins/histones<br>8. DNA unwinds into 2 strands<br>9. primer needed to start replication<br>10. DNA polymerase adds nucleotides to 3' (deoxyribose) end of strand<br>11. DNA polymerase adds nucleotides in one direction<br>12. one strand replicated continuously, the other in fragments<br>13. fragments joined by ligase | 9 |
| | B | | 1. isolation barriers prevent gene flow between populations/populations interbreeding<br>2. geographical isolation leads to allopatric speciation<br>3. behavioural isolation leads to sympatric speciation<br>4. ecological isolation leads to sympatric speciation<br>5. different mutations occur on each side of isolation barrier<br>6. some mutations may be favourable<br>7. natural selection is non-random increase in frequency of genetic sequences that increase survival<br>8. sexual selection is non-random increase in frequency of genetic sequences that increase reproductive success<br>9. Any 2 from disruptional/ directional/stabilising selection<br>10. third type of selection from 9<br>11. after many generations/long period of time<br>12. new species form<br>13. if populations can no longer interbreed to produce fertile young then different species | 9 |

# HIGHER BIOLOGY 2015

## Section 1

| Question | Answer | Mark |
|---|---|---|
| 1. | D | 1 |
| 2. | B | 1 |
| 3. | D | 1 |
| 4. | A | 1 |
| 5. | D | 1 |
| 6. | B | 1 |
| 7. | C | 1 |
| 8. | C | 1 |
| 9. | B | 1 |
| 10. | C | 1 |
| 11. | D | 1 |
| 12. | A | 1 |
| 13. | A | 1 |
| 14. | D | 1 |
| 15. | B | 1 |
| 16. | C | 1 |
| 17. | A | 1 |
| 18. | B | 1 |
| 19. | C | 1 |
| 20. | A | 1 |

## Section 2

| Question | | | Expected answer(s) | Max mark |
|---|---|---|---|---|
| 1. | (a) | (i) | ADP + Pi/phosphate/inorganic phosphate<br>*Both required for 1 mark* | 1 |
| | | (ii) | NAD | 1 |
| | | (iii) | • It is a net/overall energy gain (following an energy investment at an earlier stage)<br>**OR**<br>• More ATP/energy is produced/ released than is used/invested (earlier/in stage1) | 1 |
| | (b) | (i) | • Increases the surface area for (action of) bacteria/ Lactobacillus<br>**OR**<br>• Bursts cells to release more substrate/cell contents for bacterial action | 1 |

| Question | | | Expected answer(s) | Max mark |
|---|---|---|---|---|
| | | (ii) | • Acidic conditions/low pH/ anaerobic conditions<br>• Inhibits/kills other bacteria<br>**OR**<br>• pH/oxygen levels optimum for Lactobacillus but not for other bacteria | 1 |
| 2. | (a) | | • Same/complementary sticky ends<br>**OR**<br>• Complementary/matching base sequence/bases | 1 |
| | (b) | | (DNA) ligase | 1 |
| | (c) | | • In the presence of antibiotic only these/modified bacteria grow/survive<br>**OR**<br>• Converse | 1 |
| | (d) | | Origin of replication | 1 |
| | (e) | (i) | • Lack of post-translational modifications<br>**OR**<br>• Protein/polypeptide is folded incorrectly | 1 |
| | | (ii) | • Put modified plasmid into yeast<br>**OR**<br>• Chemically modify protein | 1 |
| 3. | (a) | (i) | Lactose concentration | 1 |
| | | (ii) | Temperature/concentration of yeast/pH | 1 |
| | (b) | | • As lactose increases (from 4%) to 16%, ethanol concentration increases 1<br>• From 16% (to 20%) ethanol remains constant/levels off 1<br>• As lactose (concentration) increases, ethanol (concentration) increases then levels off = 1 mark if neither above points included | 2 |
| | (c) | | 37.5 | 1 |
| | (d) | | • Aerobic respiration does not produce ethanol<br>**OR**<br>• Aerobic respiration produces water not ethanol<br>**OR**<br>• No/less anaerobic respiration so less ethanol produced | 1 |
| 4. | (a) | | 112 | 1 |
| | (b) | | • 08:00—12:00 1<br>• Time of lowest metabolic rate 1 | 2 |

| Question | | | Expected answer(s) | Max mark |
|---|---|---|---|---|
| | (c) | | • Energy saved/conserved<br>**OR**<br>• Uses less energy<br>**OR**<br>• Energy not wasted | 1 |
| | (d) | | • Dormancy<br>**OR**<br>• Hibernation<br>**OR**<br>• Aestivation<br>**OR**<br>• A correct description of one of these terms | 1 |
| 5. | A | | 1. Amphibian heart has 2 atria and 1 ventricle 1<br>2. Bird heart has 2 atria and 2 ventricles 1<br>3. Birds have a higher metabolic rate (or converse) 1<br>4. No mixing of oxygenated and deoxygenated blood in bird heart (or converse) 1<br>5. More efficient oxygen delivery to bird cells/muscles/tissues (or converse) 1 | 4 |
| | B | | 1. A competitive inhibitor binds to/blocks the active site 1<br>2. Competitive inhibition is reversed/reduced by increasing substrate concentration 1<br>3. Non-competitive inhibition is where a molecule binds to the enzyme not on the active site 1<br>4. Non-competitive inhibitor changes (the shape of) the active site 1<br>5. Non-competitive inhibition is irreversible/not affected by substrate concentration 1 | 4 |
| 6. | (a) | (i) | • Enzyme X: RuBisCO 1<br>• Substance Y: G3P/ glyceraldehyde-3-phosphate 1 | 2 |
| | | (ii) | • Glucose: for respiration/ ATP (production)/cellulose formation/starch formation/ other biosynthetic pathways/ processes 1<br>• RuBP: for continuation of the cycle/to allow cycle to occur/ repeat 1 | 2 |
| | (b) | (i) | • ATP: increases 1<br>• NADPH: increases 1 | 2 |

| Question | | | Expected answer(s) | Max mark |
|---|---|---|---|---|
| | | (ii) | • Procedure: randomisation of plots/treatments    1<br>• Explanation: reduces/ eliminates bias    1<br>**OR**<br>• Procedure: replication/number of replicates<br>• Explanation: to take account of variability/reduce the effect of atypical results<br>**OR**<br>• Procedure: selection of treatments/inclusion of both GM and non GM crops<br>• Explanation: to make/ensure a (fair) comparison | 2 |
| 7. | (a) | | • Parasite/Schistosoma gets energy/gets nutrients<br>**AND**<br>• Host/human is harmed (by loss of resources) | 1 |
| | (b) | | • Secondary host: Fresh water snails    1<br>• Benefit: allows development into free swimming parasite   1<br>**OR**<br>• Allows them/immature parasites to complete life cycle | 2 |
| | (c) | | • Prevent urine/faeces/eggs from entering (fresh) water<br>**OR**<br>• Stop people entering the affected water<br>**OR**<br>• Control the population of fresh water snails<br>**OR**<br>• Medication given to kill the eggs/ mature parasite/parasite in humans | 1 |
| 8. | (a) | | Biological (control) | 1 |
| | (b) | | • Harlequin ladybird has spread (rapidly)<br>**AND**<br>• Native populations/ladybirds are decreasing (and may be eliminated) | 1 |
| | (c) | | • May have alternative prey/ food source/niche/resources<br>**OR**<br>• Not competing with the Harlequin ladybird<br>**OR**<br>• Less competition with other native species<br>**OR**<br>• Not preyed upon by Harlequin ladybirds | 1 |
| | (d) | | Free from its usual predators/ parasites/pathogens/disease/ competitors | 1 |

| Question | | | Expected answer(s) | Max mark |
|---|---|---|---|---|
| 9. | (a) | (i) | 1. From 0—60 kg per hectare increase from 3—8·4 tonnes per hectare<br>2. Remains at 8.4 between 60 and 80<br>3. Between 80 and 100 decreases from 8.4—7.9/7.8 | 2 |
| | | (ii) | 208—216 | 1 |
| | | (iii) | 840 | 1 |
| | (b) | (i) | Use of 10 cattle in each group | 1 |
| | | (ii) | 0·6 | 1 |
| | | (iii) | • 20<br>• Increasing the phosphate/ fertiliser level increases the growth rate | 1<br><br>1 |
| | (c) | | • Energy lost at each level/ stage of a food chain<br>**OR**<br>• Energy lost between trophic level(s) | 1 |
| 10. | (a) | | • Horizontal    1<br>• Rapid    1 | 2 |
| | (b) | | • Resistant survive (or converse)    1<br>**AND**<br>• Transfer/pass on the (antibiotic) resistance gene/ allele to next generation    1<br>**OR**<br>• Transfer/pass on resistance gene vertically/horizontally | 2 |
| | (c) | (i) | Growth/bacteria/MRSA/colonies would be present on all plates | 1 |
| | | (ii) | Vitamins/fatty acids/beef extract | 1 |
| 11. | (a) | | 0.24 | 1 |
| | (b) | | 32 | 1 |
| | (c) | | • Inclusive scale and axes labels copied exactly from table headings    1<br>• Points plotted and joined with a ruler    1 | 2 |
| | (d) | | • Only donor 2 is suitable<br>**OR**<br>• Donor 2 is most suitable | 1 |
| | (e) | (i) | TACTGTTTAGC | 1 |
| | | (ii) | • Separates DNA strands/breaks H bonds between strands/ denatures DNA    1<br>• Any temperature from 50—65°C    1 | 2 |

| Question | | | Expected answer(s) | Max mark |
|---|---|---|---|---|
| 12. | (a) | | • Increase in stroke volume/ volume of blood pumped out of heart per heartbeat  1<br>• No effect on heart rate  1 | 2 |
| | (b) | | 1190 | 1 |
| | (c) | (i) | • Embryonic stem cells differentiate/develop into all types of cell<br>**AND**<br>• Adult/tissue stem cells differentiate/develop into narrower range of cell types<br>**OR**<br>• Adult stem cells are more differentiated/specialised than embryonic stem cells | 1 |
| | | (ii) | • They express/switch on the genes characteristic of that type of cell<br>**OR**<br>• Certain genes/some genes are expressed/switched on and other genes are switched off | 1 |
| | (d) | | Provide information on gene regulation/cell growth/cell differentiation/cell division/cell ageing/disease development | 1 |
| 13. | A | (i) | 1. RNA polymerase unzips/ unwinds DNA or separates DNA into two strands  1<br>2. Hydrogen bonds between strands/base pairs break  1<br>3. RNA polymerase aligns/ brings in/joins/attaches RNA nucleotides with their complementary nucleotides/ bases on DNA (template) or A to U and T to A and C to G in diagram  1<br>4. a primary transcript is produced  1<br>5. exons are coding and introns are non-coding (regions of the primary transcript)  1<br>6. introns/non-coding regions are removed OR exons/coding regions are retained  1<br>7. exons are spliced/joined together to form mature mRNA)/transcript  1 | 5 |

| Question | | | Expected answer(s) | Max mark |
|---|---|---|---|---|
| | | (ii) | a. tRNA has an anticodon and an amino acid attachment site  1<br>b. tRNA binds/joins to/carries/ collects specific/correct amino acid  1<br>c. tRNA carries (specific) amino acid to ribosomes  1<br>d. anticodons are complementary/pair with codons on mRNA  1<br>e. there are start and stop codons  1<br>f. peptide bonds form between amino acids  1<br>**OR**<br>a polypeptide forms | 4 |
| | B | (i) | 1. (single gene) mutations are random changes in DNA sequences/genes/alleles/the genome  1<br>2. single gene mutation name AND description<br>Substitution – base/base pair/nucleotide is replaced/ substituted by another<br>Insertion – base/base pair/ nucleotide is added/inserted<br>Deletion – base/base pair/ nucleotide is removed/ deleted  1<br>3. another single gene mutation name AND description  1<br>4. If 2 or 3 not awarded — all 3 mutation names  1<br>5. Insertion/deletion results in a frameshift mutation/expansion of a nucleic acid sequence  1<br>6. (single gene) mutations are important in evolution  1<br>7. splice site mutations can alter the mature mRNA OR result in exon removal OR result in introns remaining present  1 | 4 |

| Question | | | Expected answer(s) | Max mark |
|---|---|---|---|---|
| | | (ii) | **a.** chromosome mutation can involve changes to chromosome number/structure 1 <br> **b.** chromosome mutation name AND description; <br> Translocation: genes/sections of chromosome from one chromosome become attached to another chromosome <br> Deletion: genes/sections of chromosome deleted from chromosome <br> Inversion: genes/sections of chromosome/rotate through 1800/flipped <br> Duplication: genes/sections of chromosome/pieces of chromosome are duplicated/repeated 1 <br> **c.** another chromosome mutation name AND description 1 <br> **d.** If b or c not awarded - all 4 names but no descriptions 1 <br> **e.** polyploidy results from errors during the separation of chromosomes/non-disjunction/spindle failure during cell division/meiosis/mitosis/gamete formation 1 <br> **f.** polyploidy is the possession of complete extra sets/double/triple the number of chromosomes OR a whole genome duplication OR 2n becomes 3n 1 <br> **g.** polyploidy is important in evolution of food crops OR duplication provides material upon which natural selection can work/is important in evolution 1 <br> **h.** polyploid crops/plants/named example show desirable features/higher yields/other appropriate examples of desirable features 1 | 5 |

# HIGHER BIOLOGY 2016

## Section 1

| Question | Answer | Mark |
|---|---|---|
| 1. | B | 1 |
| 2. | A | 1 |
| 3. | C | 1 |
| 4. | B | 1 |
| 5. | C | 1 |
| 6. | B | 1 |
| 7. | A | 1 |
| 8. | A | 1 |
| 9. | C | 1 |
| 10. | C | 1 |
| 11. | A | 1 |
| 12. | D | 1 |
| 13. | A | 1 |
| 14. | D | 1 |
| 15. | D | 1 |
| 16. | B | 1 |
| 17. | D | 1 |
| 18. | C | 1 |
| 19. | B | 1 |
| 20. | D | 1 |

## Section 2

| Question | | | Expected answer(s) | Max mark |
|---|---|---|---|---|
| 1. | (a) | | Amino acid | 1 |
| | (b) | | Protein OR Enzymes | 1 |
| | (c) | | Cut/cleave **AND** combine polypeptide chains OR Add phosphate/carbohydrate | 1 |
| | (d) | | Name: **Alternative** (RNA) splicing (1) <br> Description: Different (combinations of/variety of) **exons** are included/spliced together (in the mature transcript/RNA) (1) | 2 |
| 2. | (a) | (i) | Prokaryotic has circular (chromosome) **AND** eukaryotic has linear (chromosomes) | 1 |
| | | (ii) | Proteins/Histone | 1 |

| Question | | | Expected answer(s) | Max mark |
|---|---|---|---|---|
| | (b) | | Mitochondrion OR Chloroplast OR Plasmid in yeast | 1 |
| | (c) | (i) | Nucleotides added to 3' end OR Polymerase (only) adds to 3' end OR Polymerase works from 5' to 3' OR DNA/it is replicated from 5' to 3' | 1 |
| | | (ii) | (DNA) Nucleotides/primer | 1 |
| | (d) | | So that... an exact copy/complete set of... genetic material/genetic instructions/genetic information/genes/DNA/chromosomes AND is passed to (each)... new cell/daughter cell/the next generation OR during mitosis/cell division OR So new cells have the same... genetic material/genetic instructions/genetic information/genes/DNA/chromosomes... as the original cell OR To maintain the... number of chromosomes/chromosome complement... in new/daughter cells | 1 |
| 3. | (a) | | It differentiates into/specialises into/becomes... many/lots of/all/wide range of cell types/tissue types OR It is pluripotent/totipotent | 1 |
| | (b) | | Different proteins will be produced/synthesised/made (resulting in different cell types) OR Only proteins characteristic of that cell type are produced/synthesised/made | 1 |
| | (c) | | Repair of damaged/diseased... organs/cells/tissues OR Production of tissues for grafting/transplant OR Correct examples, e.g. bone marrow transplants/(make) skin grafts/to treat a named disease/treat burns | 1 |

| Question | | | Expected answer(s) | Max mark |
|---|---|---|---|---|
| | (d) | | Embryo/it/baby/foetus/a potential life... is... destroyed/killed/not allowed to develop OR Embryos which would have been destroyed are being put to good use OR Use of stem cells for drug testing rather than animals OR Diseases could be cured | 1 |
| 4. | (a) | (i) | From... start/0 − 5 weeks/over first 5 weeks it increased from 0 − 9·2 (1) From 5 (− 7) weeks it remained constant/levelled off (1) **Correct values for 2 statements but no units (weeks) = 1 mark** | 2 |
| | | (ii) | 200 | 1 |
| | | (iii) | B | 1 |
| | (b) | | B It/number of shoots is highest/greatest (at 7 weeks) (1) and this is (still) increasing (1) OR C It/number of shoots... is increasing more/most rapidly (1) and B is slowing down/levelling off (1) | 2 |
| | (c) | | Greatest (average) root length/longer roots (1) More water absorbed for photolysis/photosynthesis OR More nutrients absorbed for named process, e.g. protein synthesis/ATP production, etc. (1) | 2 |
| 5. | (a) | (i) | Sympatric | 1 |
| | | (ii) | Prevents/interrupts/stops/blocks... gene flow/gene exchange/breeding/mating... between populations OR Prevents interbreeding | 1 |
| | | (iii) | (DNA) sequence data/genome analysis would be similar OR They/the two populations... can interbreed/breed together... to produce fertile offspring (or converse statement) | 1 |

| Question | | | Expected answer(s) | Max mark |
|---|---|---|---|---|
| | (b) | (i) | (At least one) extra set of chromosomes<br>**OR**<br>More than 2 (complete) sets of chromosomes<br>**OR**<br>2n becomes 3n/4n, etc.<br>**OR**<br>Genome duplication/multiple sets of genome | 1 |
| | | (ii) | Provides additional material upon which natural selection can work on<br>**OR**<br>Additional sets of chromosomes can mask harmful mutations<br>**OR**<br>Allows (advantageous) mutations to occur in extra chromosomes<br>**OR**<br>Can produce fertile/stable hybrids<br>**OR**<br>They are more vigorous/disease resistant/grow faster | 1 |
| 6. | (a) | | Name: Lag phase (1)<br>Explanation: (time required for) DNA replication/enzyme induction/enzyme production<br>**OR**<br>Cells can't divide until DNA replicates/enzymes induced (1)<br>**NB:** Correct explanation for lag phase with wrong name = 1 mark | 2 |
| | (b) | (i) | Stationary | 1 |
| | | (ii) | Kills/inhibits/toxic to/prevents growth of...<br>other bacteria<br>**AND**<br>reduces/eliminates competition from other bacteria<br>**OR**<br>Allows it to **outcompete** other bacteria<br>**OR**<br>Eliminates interspecific competition | 1 |
| 6. | (c) | | Cell number decreases/line goes down...<br>during/in...<br>death phase/phase D/at the end/eventually | 1 |

| Question | | | Expected answer(s) | Max mark |
|---|---|---|---|---|
| 7. | A | | 1. Anabolism is a synthesis/build up reaction<br>**OR**<br>Anabolism is build-up of molecules/substances<br>**OR**<br>Anabolism is where simple molecules are built up into more complex/large molecules | 1 |
| | | | 2. Anabolism requires the input/take up... of energy/ATP | 1 |
| | | | 3. Catabolism is breakdown/degradation...<br>of... molecules/substances<br>**OR**<br>Catabolism is a... break down/degradation... reaction<br>**OR**<br>Catabolism is where complex/large molecules are changed into more simple molecules | 1 |
| | | | 4. Energy/ATP is released/given off in catabolism | 1 |
| | | | 5. Both can have reversible and irreversible steps | 1 |
| | | | 6. Both can have alternative routes | 1 |
| | | | | **(Max 4)** |
| 7. | B | | 1. Conformers' metabolism/metabolic rate/internal environment is...<br>dependent on/affected by...<br>surroundings/external environment/external factors/external variables | 1 |
| | | | 2. Conformers use behaviour to maintain optimum **metabolic rate** | 1 |
| | | | 3. Regulators can maintain/control/regulate...<br>their...<br>metabolism/metabolic rate/internal environment...<br>regardless of external conditions | 1 |
| | | | 4. Regulators require energy for homeostasis/negative feedback | 1 |
| | | | 5. Conformers have **narrower** (ecological) niches (or converse) | 1 |
| | | | 6. Conformers have low**er** metabolic costs/rates of metabolism<br>(or converse) | 1 |
| | | | | **(Max 4)** |

| Question | | | Expected answer(s) | Max mark |
|---|---|---|---|---|
| **8.** | (a) | | 990 | 1 |
| | (b) | | As temperature increases population decreases **OR** The higher the temperature the lower the population **NB:** If values included (21 to 72)/(123 to 0·1) they must be correct, units not necessary **NB:** Any description extended beyond the first 4 days negates | 1 |
| | (c) | (i) | Species: B **(1)** Justification: high population/population thrived at… 72°C/highest temperature **OR** higher population than A or C at… 72°C/highest temperature **(1)** | 2 |
| | | (ii) | Contain enzymes/proteins which are… tolerant of/don't denature at/are resistant to/optimum at/ working at… high temperatures | 1 |
| | | (iii) | hot springs/geysers/volcanoes/ seabed vents | 1 |
| **9.** | (a) | | Name: (restriction) endonuclease **(1)** Function: Cuts DNA/genes out **OR** Cuts plasmid **(1)** **OR** Name: Ligase **(1)** Function: Joins/seals/inserts gene **into** plasmid **OR** Joins/seals sticky ends of plasmid and gene **(1)** | 2 |
| | (b) | (i) | Grow/culture with ampicillin/antibiotic **(1)** Only cells containing the plasmid/that gene/transformed cells/modified cells/can grow/ survive **(1)** | 2 |
| | | (ii) | DNA/gene/plasmid/genetic info. passed from/between/to… cell/bacterium/bacteria… in same generation/without reproduction/in same population/neighbouring bacteria **OR** DNA/gene/plasmid/genetic information/vector passed by… conjugation/transduction/ transformation (or description) **OR** DNA/gene/plasmid/genetic info. passed from prokaryote to eukaryote | 1 |

| Question | | | Expected answer(s) | Max mark |
|---|---|---|---|---|
| | (c) | | Eliminates/kills… other/contaminating/ unwanted… micro-organisms/ bacteria **OR** Eliminates competition from… other/unwanted… micro-organisms/bacteria **OR** So **only** insulin-producing bacteria can grow | 1 |
| **10.** | (a) | (i) | To allow (time) for… respiration/metabolic rate… to be affected by… temperature/conditions/change **OR** To allow crickets (time) to… acclimatise/adjust/respond to/ get used to… temperature/condition/change **OR** To allow flask/equipment/ crickets (time) to reach the temperature | 1 |
| | | (ii) | Description: (exactly) the same… set up/experiment… **OR** Full description (same size/ volume of flask, in water bath and $CO_2$ sensor) **AND** (with) no crickets/dead crickets/ glass beads **(1)** Explanation: To show it was the crickets that respired/metabolised/produced the $CO_2$ **OR** No… $CO_2$ production/respiration/ metabolism… without live crickets/with dead crickets/with no crickets/with control **(1)** | 2 |
| | (b) | | Axes labelled correctly and scales to fill at least half the grid **(1)** Points plotted correctly and joined with a ruler **(1)** | 2 |
| | (c) | | As the temperature increased, the (rate of) metabolism increased | 1 |

| Question | | | Expected answer(s) | Max mark |
|---|---|---|---|---|
| **11.** | (a) | (i) | Colchicine concentration | 1 |
| | | (ii) | 50 plants/seeds **at each concentration** | 1 |
| | (b) | (i) | 8 | 1 |
| | | (ii) | 3 : 7 | 1 |
| | (c) | | More photosynthesis (1)<br>More energy for growth/seed production (1) | 2 |
| **12.** | (a) | (i) | (Female) mosquito | 1 |
| | | (ii) | Females/they need the **blood** for egg **production**<br>**OR**<br>Males don't produce eggs so don't need **blood** | 1 |
| | (b) | | (The host is harmed) by losing energy/nutrients/food<br>**OR**<br>(Host harmed as) parasite feeds off it/gains nutrients from it | 1 |
| | (c) | | Method 1:<br>Mosquito… discouraged/stopped from… biting/feeding/fewer people bitten<br>**AND**<br>it cannot spread parasite/disease/virus/bacteria<br>**OR**<br>Method 2:<br>There are… no/fewer… parasites to transmit to the human/mosquito (1) | 1 |
| **13.** | (a) | (i) | 110 | 1 |
| | | (ii) | 3100 | 1 |
| | | (iii) | 325 | 1 |
| | (b) | | Zebra mussel population increased and unionid decreased | 1 |
| | (c) | | Unionid/native population drops (from 140) to zero/killed off/eliminated | 1 |

| Question | | | Expected answer(s) | Max mark |
|---|---|---|---|---|
| | (d) | | New environment may be free from/have less/have no…<br>predators<br>**OR** parasite/disease<br>**OR** pathogens<br>**OR** competitors<br>(which would limit its population in its native habitat) | 1 |
| | (e) | | Number/abundance<br>**AND**<br>frequency of alleles in a population/gene pool/species | 1 |
| **14.** | A | (i) | Weeds:<br>1. Weeds compete with/inhibit (crop) plants<br>**AND**<br>reduce productivity/growth/yield | 1 |
| | | | 2. Annual weeds have…<br>rapid growth/short life cycles/complete life cycle within a year/produce many seeds/produce seeds with long term viability<br>**OR**<br>Perennial weeds have storage organs/vegetative reproduction | 1 |
| | | | Pests:<br>3. Pests eat/damage… crops/plants/plant parts<br>**AND**<br>reduce productivity/growth/yield | 1 |
| | | | 4. Any 2 from nematodes/insects/molluscs | 1 |
| | | | Diseases:<br>5. Diseases are caused by bacteria/fungi/viruses | 1 |
| | | | 6. Diseases are often spread by invertebrates/pests | 1 |
| | | | | (Max 4) |

| Question | | | Expected answer(s) | Max mark |
|---|---|---|---|---|
| 14. | A | (ii) | a. Weeds/pests/diseases can be controlled by <u>cultural</u> means **AND** example (ploughing/weeding/roguing/crop rotation/time of sowing) | 1 |
| | | | b. Selective weed killer/selective herbicide... only kills/affects... certain plant species/broad leaved weed | 1 |
| | | | c. Systemic weed killer... spreads through (vascular system)/enters plant **AND** kills (whole) plant/stops regrowth/regeneration **OR** Systemic pesticide/insecticide... spreads through/enters plant/in the phloem **AND** kills pest feeding on plant | 1 |
| | | | d. Applications of fungicide based on disease forecasts are more effective than treating diseased crop | 1 |
| | | | e. Compensatory mark to be awarded if none of points b/c/d awarded but fungicide, pesticide/insecticide and herbicide are all named | 1 |
| | | | f. Biological control is use of predator/parasite of pest | 1 |
| | | | g. Example of a problem with chemical/biological control ANY <u>one</u> from:<br>• toxicity to non-pest/target species<br>• persistence in the environment<br>• accumulation/magnification in food chains/food webs<br>• production of resistant populations<br>• predator/parasite/control organism/disease organism becomes... invasive/outcompetes/preys on/parasitizes other species | 1 |
| | | | h. Integrated pest management combines biological and chemical control **OR** chemical and cultural control **OR** biological and cultural control **OR** biological, cultural and chemical control | 1 |
| | | | | (Max 4) |

| Question | | | Expected answer(s) | Max mark |
|---|---|---|---|---|
| 14. | B | (i) | 1. Social hierarchy is a rank order/pecking order in a group of animals **OR** Dominant/alpha **AND** subordinates/lower rank | 1 |
| | | | 2. Aggression/fighting/conflict/violence reduced | 1 |
| | | | 3. Ritualistic display/appeasement/threat/submissive... behaviour **OR** Alliances formed to increase social status **OR** Descriptive examples | 1 |
| | | | 4. Ensures... best/successful... genes/characteristics are passed on **OR** Guarantees experienced leadership | 1 |
| | | | **Max of 3 marks from points 1 to 4** | |
| | | | 5. Cooperative hunting is where animals hunt in a group/together **AND one** from: Increases hunting success **OR** Allows larger prey to be brought down **OR** More successful than hunting individually | 1 |
| | | | 6. (Subordinate) animals all get **more** food/energy... than **hunting alone** | 1 |
| | | | 7. Less energy used/lost per individual | 1 |
| | | | | (Max 4) |

| Question | | | Expected answer(s) | Max mark |
|---|---|---|---|---|
| | | (ii) | **a.** Any 2 examples – bees, wasps, ants, termites | 1 |
| | | | **b.** Only some members of colony (hive) reproduce/are fertile<br>**OR**<br>Queen **AND** males/drones mate/reproduce<br>**OR**<br>Only queen lays eggs<br>**OR**<br>Some/most members of colony are sterile/infertile/do not reproduce<br>**OR**<br>Some/most of colony are workers who are sterile | 1 |
| | | | **c.** Examples of worker roles ANY **one** from:<br>• raise relatives<br>• defend the hive<br>• collect pollen/nectar/food<br>• waggle dance to show direction of food, etc. | 1 |
| | | | **d.** Social insects show...<br>kin selection/altruism between related individuals | 1 |
| | | | **e.** Increases/helps survival of shared genes<br>**OR**<br>So shared genes are passed on to next generation | 1 |
| | | | **f.** Some are **keystone species** which are crucial for (stability of) the ecosystem/food web/pollination/soil fertility<br>**OR**<br>The removal of **keystone species** can...<br>disrupt/collapse...<br>the ecosystem/food web | 1 |
| | | | | (Max 4) |

# Acknowledgements

Permission has been sought from all relevant copyright holders and Hodder Gibson is grateful for the use of the following:

Image © BlueRingMedia/Shutterstock.com (2016 Paper 2 page 8).